23rd International Conference on Vacuum Technique and Technology 2016

Journal of Physics: Conference Series
Volume 729

St. Petersburg, Russia
7 – 9 June 2016

Editors:

A.A. Lisenkov
D.K. Kostrin
V.A. Pavlova

ISBN: 978-1-5108-2727-1
ISSN: 1742-6588

Printed from e-media with permission by:

Curran Associates, Inc.
57 Morehouse Lane
Red Hook, NY 12571

Some format issues inherent in the e-media version may also appear in this print version.

Printed by Curran Associates, Inc. (2016)

For permission requests, please contact the Institute of Physics at the address below.

Institute of Physics
Dirac House, Temple Back
Bristol BS1 6BE UK

Phone: 44 1 17 929 7481
Fax: 44 1 17 920 0979

techtracking@iop.org

Additional copies of this publication are available from:

Curran Associates, Inc.
57 Morehouse Lane
Red Hook, NY 12571 USA
Phone: 845-758-0400
Fax: 845-758-2633
Email: curran@proceedings.com
Web: www.proceedings.com

Table of contents

Volume 729

23rd International Conference on Vacuum Technique and Technology
7–9 June 2016, St. Petersburg, Russia

Accepted papers received: 29 June 2016
Published online: 25 July 2016

Aims and Results of the 23rd International Conference on Vacuum Technique and Technology (VTT2016)

A A Lisenkov[1], **D K Kostrin**[2] **and V A Pavlova**[3]

[1] Institute of Problems of Mechanical Engineering, Russian Academy of Sciences, 199178, Saint Petersburg, Russia
[2] Saint Petersburg Electrotechnical University “LETI”, 197376, Saint Petersburg, Russia
[3] Saint Petersburg Polytechnic University, 195251, Saint Petersburg, Russia

E-mail: lisran@yandex.ru

Abstract. In this preface the main features, aims and results of the 23rd International Conference on Vacuum Technique and Technology (VTT2016) that was held on 7–9 June 2016 in Saint Petersburg, Russia are discussed.

The International Conference on Vacuum Technique and Technology is held annually, the main organizers of the event are (figure 1):

- Russian Academy of Sciences (RAS);
- Institute of Problems of Mechanical Engineering RAS (http://www.ipme.ru);
- Saint Petersburg Polytechnic University (http://english.spbstu.ru);
- Saint Petersburg Electrotechnical University "LETI" (http://eltech.ru/en/university).

For the last several years the conference is held at Saint Petersburg Electrotechnical University "LETI" (figure 2) – alma mater of the 2000 Nobel Prize winner in Physics, Zhores Alferov.

Figure 1. Logos of the organizers of the 23rd International Conference on Vacuum Technique and Technology (VTT2016).

The members of organizing and scientific committees were the leading Russian scientists:

- Afanasjev V P – Full Prof., D. Sc., Saint Petersburg Electrotechnical University "LETI";
- Koval N N – Full Prof., D. Sc., Institute of High Current Electronics RAS;

- Kuznetsov V G – Full Prof., D. Sc., Institute of Problems of Mechanical Engineering RAS;
- Lisenkov A A – Full Prof., D. Sc., Institute of Problems of Mechanical Engineering RAS;
- Luchinin V V – Full Prof., D. Sc., Saint Petersburg Electrotechnical University "LETI";
- Lozovan A A – Full Prof., D. Sc., Russian State Technological University MATI;
- Nesterov S B – Full Prof., D. Sc., Scientific Research Institute of Vacuum Technologies;
- Potrakhov N N – Full Prof., D. Sc., Saint Petersburg Electrotechnical University "LETI";
- Rozanov L N – Full Prof., D. Sc., Saint Petersburg Polytechnic University;
- Saksaganskiy G L – Full Prof., D. Sc., Efremov Institute of Electrophysical Apparatus;
- Frolov V Ya – Full Prof., D. Sc., Saint Petersburg Polytechnic University.

Figure 2. Saint Petersburg Electrotechnical University "LETI" (founded in 1886).

The event was held in several sections:

- Vacuum technique, vacuum measurement and mass spectrometry;
- Vacuum ion-plasma methods of surface treatment;
- Nanomaterials and nanotechnology;
- Physical processes at all stages of coating formation;
- Technology and properties of coatings and films and methods of their analysis.

The aims of the conference were:

- to present new achievements and results of theoretical and experimental research in the physics of vacuums and vacuum measurement, mass spectrometry and leak testing;
- to consider topical issues of producing vacuums, use of elements of vacuum systems and creation of new vacuum equipment;
- to discuss the application of vacuum technology in surface modification and formation of thin films and coatings using ion-plasma and beam technologies;
- to assess the prospects of applying ion-plasma technologies for synthesis of new materials, including nanomaterials;
- to facilitate the integration of efforts of research teams through discussion of new ideas and results, and testing of innovative practical developments.

To achieve these objectives the following activities were carried out:

- exchange of experience and fruitful discussions on the development and operation of vacuum equipment, new technological devices, technologies, materials and methods of research;
- presentation of new developments of equipment, components and parts, technologies, control systems, vacuum and automation elements;

- development of relations and establishing of new contacts between industry and science, manufacturers and developers of technical management systems, suppliers and customers of the vacuum equipment;
- stimulation of the creative potential of young scientists in the field of vacuum and vacuum technology.

The following participated in the 23rd International Conference on Vacuum Technique and Technology (VTT2016):

- institutes of the Academies of Sciences of Russia, Belarus and Ukraine;
- leading higher education institutions;
- scientific research institutes;
- production companies.

The total number of participants was 127 (figure 3), representing various regions of Russia and also Belarus, Ukraine, Germany and Vietnam. During the Conference 84 oral and poster reports involving all areas of vacuum technique and technology were presented. The best 30 papers were chosen by the Scientific Committee to be presented in this volume of Journal of Physics: Conference Series.

Figure 3. Participants of the second day of the 23rd International Conference on Vacuum Technique and Technology (VTT2016).

All presentations by the participants of the Conference inspired the scientific interest of the audience. In the general discussion, questions, comments and opinions were presented. It should be noted and appreciated that a high professionalism of the speakers and also high ethics of communication were present among the participants during the exchange of views.

In order to summarize the results of the Conference its participants noted that the diversity and number of submitted reports was a sign of the continuous technical development of the industry, and responsibility of the participant groups for its future development.

Journal of Physics: Conference Series **729** (2016) 011002 doi:10.1088/1742-6596/729/1/011002

Peer review statement

All papers published in this volume of *Journal of Physics: Conference Series* have been peer reviewed through processes administered by the proceedings Editors. Reviews were conducted by expert referees to the professional and scientific standards expected of a proceedings journal published by IOP Publishing.

The water outgassing rate of internal surfaces of vacuum systems

L N Rozanov

St. Petersburg Polytechnic University, 195251, St. Petersburg, Polytechnicheskaya st. 29, Russia

E-mail: lnrozanov@mail.ru

Abstract. On the basis of experimental adsorption isotherm the ratio between the real and geometrical surfaces was calculated and the amount of gas required to form a monolayer was defined. Simultaneous usage of Henry and Frendlih equations allowed to determine the dependence of the heat of adsorption on the logarithm of the absorbed gas amount A mathematical model of pumping of the vacuum systems with adsorbing walls is presented. This model uses the parameters of the vacuum system and the dependence of the adsorption heat on the amount of the adsorbed gas .The conditions of the existence of regular pumping regime are discussed. The structure database vacuum adsorption properties of materials was proposed. The experimental data on the determination of the adsorption outgassing rate were released.

1. Introduction

Adsorption outgassing from internal surfaces of vacuum systems is the main component of the gas load in unheated vacuum systems. It is associated with desorption of water vapor adsorbed from the atmosphere of vacuum chamber during depressurization. There are diffusion [1] and the adsorption [2] models of outgassing in vacuum systems. This work is devoted to the discussion of the adsorption model. The work [2] is proposed to use the Temkin's isotherm equation with the linear dependence of the heat of adsorption on the amount of adsorbed gas.

This work describes adsorption model based on Frendlih's isotherm equation that describe the experimental curves well. Experimental curves of outgassing published in the works [1–3, 9] for the same materials, differ within an order of magnitude and are suitable for the calculation of gasload for vacuum systems, whose parameters are similar to the experimental units, in which the coefficients of outgassing were determined. In this work for calculating of outgassing rates we are taken into account the gas adsorption properties of the surface and the parameters of the vacuum system. This model is the base of method of forming the database to determine the rates of adsorption outgassing of different vacuum settings.

2. The composition of the surface

The basic material for manufacturing of vacuum chambers is a non-magnetic austenitic stainless steel. Aluminum alloys are used less and titanium alloys are used seldom. The elemental and quantitative composition of the stainless 12X18H10 surface before and after ion etching was analyzed by the photoelectron spectrometer Thermofisher Scientific Escalab 250Xi .The parameters of first etching cycle of surface were: energy of argon ions 3000 eV, spot's diameter: 1 mm, etching time: 10 s, the angle between the surface normal and the direction of the incident ions – 40°; in second cycle of etching the time was increased to 60 s. The results of the elemental composition of the surface are shown in table 1.

Table 1. Elemental composition of the surface.

Element	% of atoms		
	Before etching	After 1 cycle of etching	After 2 cycles of etching
C	48.49	–	–
O	44.42	26.79	5.80
Cr	4.76	19.87	20.07
Ni	–	6.89	7.35
Ca	1.51	–	–
Fe	0.82	46.44	66.78

A large amount of carbon before the first etching cycle indicates sample contamination. The composition of the main component of steel 12X18H10T in table after the second cycle of etching, in general, coincides with the volumetric composition of the steel. A noticeable amount of oxygen on the surface may indicate the oxide presence. The results imply that for untreated stainless steel the adsorption of water vapor occurs on the surface polluted by carbon. After cleaning the surface adsorption takes place largely on the oxides. The results are confirmed by the data in work [9].

3. Adsorption isotherms

The adsorption isotherm on untreated stainless steel (R_a = 17 μm, T = 298 K) in the pressure range $1\ldots10^3$ Pa determined by volumetric method as in work [6] is well described by Frendlih equation:

$$a(p) = 1.85\cdot10^{-2}p^{0.34}, \tag{1}$$

where a – the specific amount of adsorbed gas, m^3Pa/m^2; p – pressure of water vapor, Pa.

The adsorption isotherm on electropolished stainless steel at T = 310 K determined by dynamic method [4], in the pressure range of 10^{-2}–10^{-8} Pa is well-described by Frendlih equation:

$$a(p) = 1.76\cdot10^{-2}p^{0.25}. \tag{2}$$

Electropolishing leads to decreasing the pressure exponent.

4. The surface structure

The real surface of the vacuum chamber is not smooth. It has roughness, cracks and other defects. The relation between the real and the geometric surface may be determined using BET adsorption isotherm [5]. Considering that at the smooth surface can be placed $4.9\cdot10^{18}$ $1/m^2$ of water molecules, we found that the real surface of the raw stainless steel is bigger in 5.8 times than geometric surface. We assume that the water condensation probability $s = 1$.

5. The heat of adsorption

Henry and Frendlih isotherm equation gave similar results over a wide range of pressures if a linear dependence of the heat of adsorption on the logarithm of the amount of absorbed gas exists:

$$Q_a = Q_1 - K_Q\,\lg(a)\,, \tag{3}$$

where $Q_1 = RT\ln\left(\dfrac{d^{1/m}}{K\cdot\tau_0}\right)$; $K_Q = \dfrac{RT(1-m)}{0.43\cdot m}$.

Journal of Physics: Conference Series **729** (2016) 012001 doi:10.1088/1742-6596/729/1/012001

According to equation (1) for water on the stainless steel: $d = 0.0185$; $m = 0.34$. At $\tau_0 = 1.6\cdot 10^{-13}$ s and $K = 147.5$ we receive that $Q_1 = 3.15\cdot 10^7$ J/kmol and $K_Q = 1.12\cdot 10^7$ J/kmol.

6. The model of pumping of vacuum chamber with adsorbing walls

Without pumping of vacuum chamber ($q_p = 0$) (figure 1) processes of evaporation q_e and condensation q_c of gas molecules occur at the free surface simultaneously.

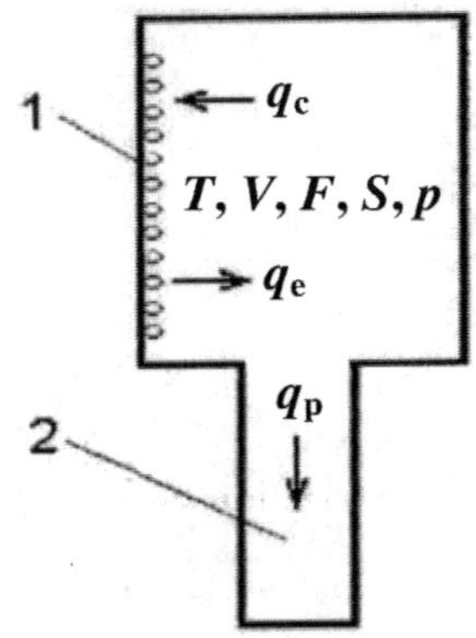

Figure 1. The model of pumping of the vacuum chamber with absorbing walls: 1 – vacuum chamber; 2 – the vacuum pump.

For pumping water vapor at 298 K in vacuum systems equation $a = dp^m$ of Frendlih's adsorption isotherm can be used. Adsorption outgassing rate will be determined by the formula

$$q_a = -\frac{da}{dp}\cdot\frac{dp}{dt} = \frac{d\cdot m\cdot S\cdot p^m}{V + F\cdot d\cdot m\cdot p^{m-1}}. \quad (4)$$

The model of pumping vacuum chamber with adsorbing walls can be presented:

$$a = d\cdot p^m; \quad \frac{dp}{dt} + \frac{S\cdot p}{V + V_\partial} = 0;$$

$$V_d = F\cdot m\cdot d\cdot p^{m-1}; \quad q_a = \frac{d\cdot m\cdot S\cdot p^m}{V + F\cdot d\cdot m\cdot p^{m-1}}; \quad (5)$$

$$S = K_e\cdot S_m\left(1 - \frac{p_m}{K_e p}\right).$$

The results of calculations on the (5) at $T = 298$ K; $M = 18$ kg/kmol; $s = 1$; $d = 0.0185$; $m = 0.34$; $V = 1$ m^3; $F = 10$ m^2; $S_m = 0.2$ m^3/s; $p_f = 10^{-8}$ Pa; $p_s = 10^3$ Pa and $K_e = 1$ give the graphs on the time (figure 2).

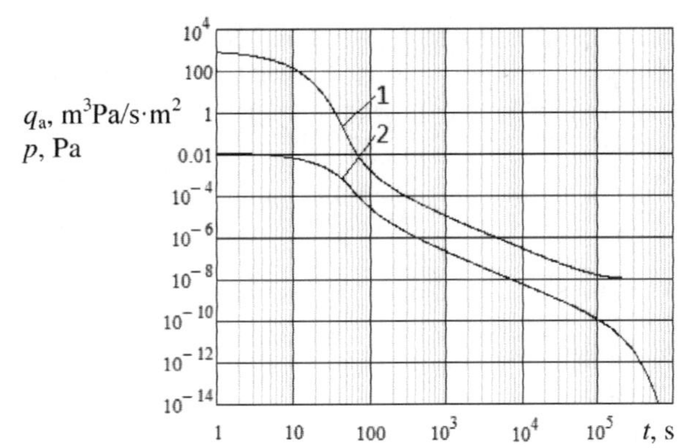

Figure 2. Dependence of pressure (1) and outgassing rate (2) of vacuum system on pumping time.

7. Dependence of the adsorption outgassing rate on pumping time and parameters of vacuum system

Adsorption outgassing rate during pumping of the vacuum chamber in accordance with equations (5) depends on coefficients of the adsorption isotherm and parameters of the vacuum system. The equations for regular mode depend only on the surface area and the effective pumping speed of chamber. They can be obtained by integrating (4) and neglecting the factors depending on the volume of the chamber, initial and ultimate pressures. Regular mode in figure 2 is executed at pumping times in range 10^2–10^5 s.The dependence of the pressure on the pumping time in the regular mode is as follows:

$$p(t)=\left[\frac{d\cdot m\cdot F}{S(1-m)t}\right]^{\frac{1}{1-m}}. \qquad (6)$$

The dependence of the adsorption outgassing rate on pumping time in regular mode has the form:

$$q_a(t)=\frac{S\cdot p(t)}{F}=\left(\frac{F}{S}\right)^{\frac{m}{1-m}}\left[\frac{d\cdot m}{(1-m)t}\right]^{\frac{1}{1-m}}. \qquad (7)$$

8. The database of adsorption properties of vacuum materials

The database of adsorption properties of vacuum materials may be composed of the coefficients *d* and *m* of Frendlih's isotherm equation (table 2).

Table 2. The adsorption outgassing rates of water vapor at T = 298 K.

Material	Preliminary processing	d,[m^{3}Pa]m^{-2}Pa^{-m}	m	Reference
Stainless steel X18H9T	Without processing	0.0185	0.34	This work
	Electropolishing	0.0176	0.25	[4]

The calculations in regular mode (7) except *d* and *m* from table 2 should be completed by the parameters of vacuum system (F, S).

For example, when F_c = 10 m^2, S = 0.2 m^3/s after 1 hour of pumping the specific outgassing rate according the equation (7) is equal to $2.3\cdot10^{-7}$ m^{3}Pa/(s·m^2). The value m = 0.34 (table 2) is confirmed by our experiments and works [1, 9].

5. Conclusions

Untreated stainless steel surface is polluted by carbon. Real surface is 5.8 times bigger than geometrical one. Water adsorption isotherm is described by Frendlih equation with parameters d = 0.0185 and m = 0.34. The proposed mathematical model in regular regime complies experimental results.

References

[1] Li M and Dylla H F 1994 *JVST A* **4** 1772–7
[2] Redhead P A 1995 *JVST A* **2** 467–75
[3] Nuss H E 1996 *Vacuum* **4** 391–5
[4] Jilnin V S, Jilnina L N and Kuzmin A A 1974 *Electronic Engineering. Ser. Electric vacuum and gas appliances* **4** 78–80
[5] Rozanov L N 1994 *Vacuum technique and technology* **4** 24–6
[6] Rozanov L N and Akimov Y D 1985 *Journal of Applied Chemistry* **6** 1381–4

Journal of Physics: Conference Series **729** (2016) 012001 doi:10.1088/1742-6596/729/1/012001

[7] Rozanov L N 2002 *Vacuum Technique* (London, New York: Taylor and Francis)
[8] Garker B, Edellman Chr and Ehert M 1996 *Vacuum* **4** 383–90
[9] Chen J R, Huang J R, Hsiung G Y, Wu T Y and Liu Y C 1994 *JVST A* **4** 1750–4

Results of mathematical modelling the kinetics of gaseous exchange through small channels in micro dischargers

S A Bushin

Dukhov Research Institute of Automatics, 22 Suschevskaya St., Moscow, Russia

E-mail: vniia4@vniia.ru

Abstract. Results obtained using mathematical calculating models for physical processes of gaseous exchange through low-conductivity channels in the sealed envelopes of dischargers for various flow modes of indicative working gas are presented.

Quantitative assessment of leakage flows (inleakage) while mass spectrometric monitoring of leak tightness is still a topical challenge in spite of available computerized leak detectors with the implemented initial data processing algorithms. In most cases the assessment is made through comparing the unknown values of flows with the known ones for the standards. At the same time the calibrated leaks used as the gas flow measures (such as of GELIT type) correspond with higher degree of confidence to real gas exchange conditions through leaks in case of molecular flow mode, and with less confidence in case of viscosity mode.

As a rule, obtaining of reliable results of flow rates through open-end channels is problematic because of the lack of data on dimensions and shapes of the controlled leaks, therefore it is quite a laborious task to get the result approximating to the actual value without using special methods and calculations.

The aim of this paper is to analyze and investigate the dynamics of working gas mass flow rate (leak) through the leakages in gas-filled micro dischargers depending on determined effect of different gas flow modes taking into account specified sequence of the process operations while high-sensitive monitoring of leak tightness [1]. In case of envelope leaks, the leak (inleakage) flow rates depend in a certain way on alternate transitional processes caused by the change of pressure inside the recipients due to variation of external conditions (keeping in an open air after production, evacuation, pressure testing (in case of large leaks), the following repeated evacuation, discharge of the working gas to vacuum (without evacuation)) that are a part of operations accompanying the control. Besides it was necessary to find certain interrelation between the volume of gaseous neon in the atmosphere and a possibility to extend essentially operating range of the detected comparatively large leaks by means of increasing the upper sensitivity threshold for rejecting the dischargers using mentioned gas as a working medium.

It is difficult to apply leak detection equations for calculation since it is impossible to examine small channels visually and measure them; and their geometric parameters cannot be used as initial data as it is made for vacuum calculations of gas flows Q through pipelines. Therefore the most rational approach to the calculation-based leak detection is to use relative assessment. When the flow ratios for different substances going through the same leak at different pressure values but within the same flow mode are calculated, the parameters describing the leak geometry are excluded. An air flow

B from atmosphere to vacuum at room temperature, i. e. gas flow in normalized conditions is taken as initial characteristic of leak value. However confidence of calculation is limited by uncertainty of boundaries of purely viscous and molecular modes, when there is no data, as noted above, on leak channel dimensions [2].

Based on the differential form of the universal empirical equation proposed [3] for describing gaseous flows through leaks of arbitrary geometry, the results of theoretical investigations [4, 5] on assessment of neon leak flows from gas-filled micro dischargers ($V_p = 0.43$ cm^3) through leakages were obtained for general case of viscous-molecular-diffusion flow mode. Calculations were made using relative parameter φ ($\varphi = B_d/B_n$, where B_d is a diffusion leakage defined by a neon flow from the gas atmosphere to air (if there is no total pressure differential), B_n is a neon leakage characterized by the neon flow from the gas atmosphere to vacuum). It should be noted that the nondimensional parameter introduced fully characterizes the flow mode from viscosity $\varphi = 0$ to molecular mode $\varphi = 1$ [6].

Theoretical calculation research for different working gas flow modes (molecular, diffusion and viscous) was based on calculation of controlled flow Q_n as a function of the leak rate *B* ($Q_n = f(B)$) with consideration of atmospheric neon component and without it. Q_n was calculated using the systems of mathematical models that describe approximately specified combinations from the sequence of the standard operations executed as stated in the developed methodology for the finishing control of the leak tightness implemented on computerized mass-spectrometer setup UFKG [7]. However it is impossible to show all the results obtained in this work, so it is suggested to consider that investigations were carried out in two stages in order to limit the discussion (figure 1).

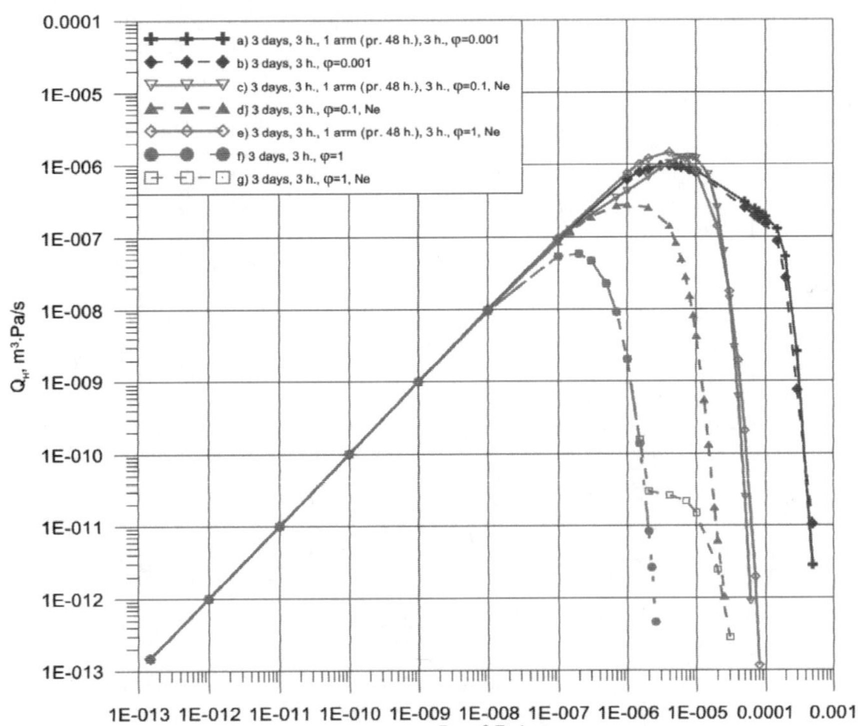

Figure 1. Superposition of curves that show $Q_n = f(B)$ functions obtained at stages 1 and 2 (table 1). Curves a, b, f and c, d, e, g are plotted respectively with and without consideration of neon in atmosphere at $\varphi = 0.001$ (a, b); $\varphi = 0.1$ (c, d); $\varphi = 1$ (e, f, g).

The first stage included calculations with analytical consideration of storage of the discharger with hypothetic leak B_i of the specified levels in the air; time was reckoned starting from the discharger manufacture (after it was filled with the neon working gas at pressure $p_n = P_a^*$ (101.3 kPa here) and when the exhaust tube was sealed) for time period T_{vd} with the following evacuation lasted for t_{ev}. The

Journal of Physics: Conference Series **729** (2016) 012002 doi:10.1088/1742-6596/729/1/012002

second stage of investigations consisted of all the operations from the first stage, but there were additionally introduced by mathematical simulation: neon pressure test at P_{pr} for T_{pr} and the following repeated evacuation for t_{ev} (table1).

Table 1. Parameters of the investigation stages.

Stage	T_{vd}, h	P_{pr}, atm	T_{pr}, h	t_{ev}, h
1	72	0	0	3
2	72	1	48	3

Figure 1 demonstrates graphical view of the obtained theoretical functions $Q_n = f(B)$ that show that the most pronounced effect of atmospheric neon penetrating through the leak(s) into the sealed envelope of the discharger chamber takes place in the molecular flow mode. And it is possible to explain a spike in the bottom of g-curve (3 d, 3 h, φ = 1, Ne (see figure 1)) by means of analyzing the functions from figure 2.

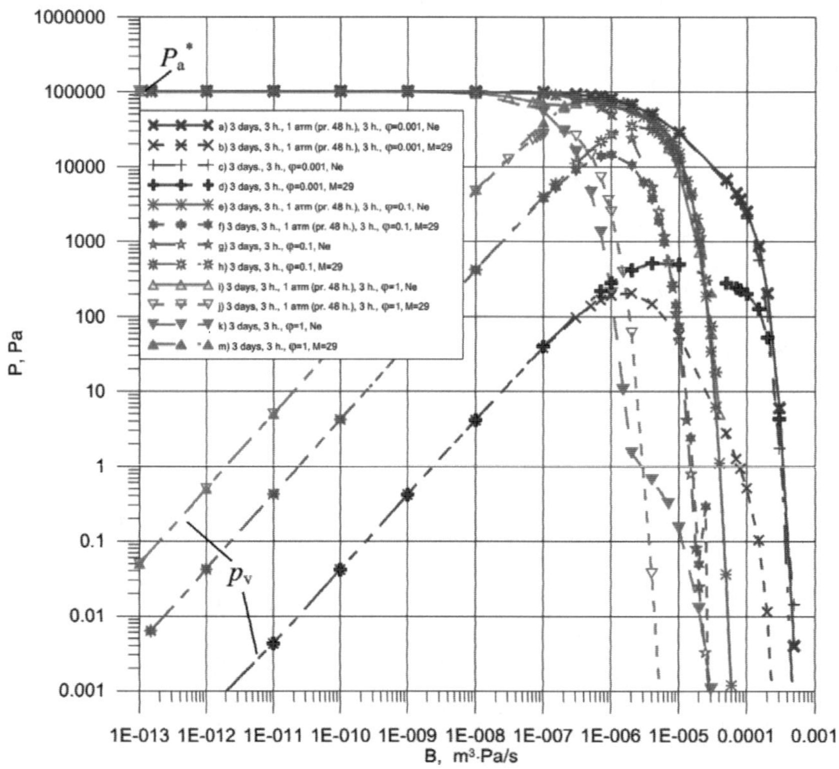

Figure 2. Superposition of functions that show the change of neon pressure $p_n = f(B)$ and air pressure $p_v = f(B)$ in the discharger after the end of control operations for two research stages (table 1). Curves a, c, e, g, i, k correspond to the atmospheric neon gas taking into account its presence in the atmosphere; curves b, d, f, j, l, m are for air at φ = 0.001; φ = 0.1; φ = 1 respectively.

The analysis of the results (figures 1, 2) obtained for viscosity mode (φ = 0.001) show that while discharger storage in the air the partial pressure of the neon gas inside even at the leak values B up to $\approx 10^{-6}$ Pa·m^3/s is kept at the level $p_n \approx P_a^*$. Pressure test in the neon medium has no significant effect on the neon partial pressure in the same way as for the discharger that was not subject to pressure testing. This, in its turn, consequently has no effect on the leak rates, though the values of the leakage are quite high. The repeated evacuation of the working volume of the down-hole chamber with the discharger before monitoring also has practically no effect on leak rates, however there is a sharp fall within an open interval $B \geq 2\cdot10^{-4}$ Pa·m^3/s on the curves for the neon flow $Q_n = f(B)$ (figure 1, curves a, b). The dynamics of air pressure p_a inside the instrument remains quite stable within the leak range from $2\cdot10^{-12}$ to $3\cdot10^{-7}$ Pa·m^3/s, and the repeated evacuation after the pressure test starts influencing significantly on the air pressure level for leak rates $B \geq 5\cdot10^{-6}$ Pa·m^3/s (figure 2, curve b), and the pressure is going down practically by six decades (10^6) unlike similar version of investigations 1 (figure 2, curve d) without pressure test and repeated evacuation, where p_a is ranged between 150 to 500 Pa.

For the molecular flow mode ($\varphi = 1$) with leaks B from 10^{-7} to 10^{-6} Pa·m^3/s in the dischargers the partial neon pressure after pressure test is approximately equal to the molding pressure. The neon flow Q_n is slowly going up in the area of leaks $B = (5\cdot10^{-7}...5\cdot10^{-6})$ Pa·m^3/s due to the increase of conductivity of the leak channel B; and increase of its numeric value influences on Q_n value here greater than little decrease of neon pressure p_n in the discharger.

The neon pressure difference in case of $B < 10^{-7}$ Pa·m^3/s leaks is practically similar for viscosity and molecular flow modes, but more significant neon loss in molecular flow mode initiates fall of Q_n on the curve $Q_n = f(B)$ at essentially less values of Q_n than in case of $\varphi = 0$. Tending to slow fall of Q_n at $B > 2\cdot10^{-6}$ Pa·m^3/s (figure 1, curve g and figure 2, curve k) relates to the area of neon pressure p_n in the discharger due to its equalizing while keeping in the air to the level comparable with the partial pressure of the atmospheric neon (1.8 Pa). It is evident that the discussed area characterizes capabilities of rejecting gas-filled dischargers with the specified storage time based on the content of atmospheric neon inside them without additional pressure test. Hence for research stage 1 this effect can be used in practice for the extension of the upper check limit of dischargers rejection from $3\cdot10^{-6}$ to $3\cdot10^{-5}$ Pa·m^3/s.

Thus based on the analytical assessment of gas exchange through the open-end low-conductivity channels in the sealed envelopes of gas-filled micro dischargers it is determined that the values of relatively large leaks depend on the working gas flow through the leaks as well as on a sequence of the process operations performed during control considering a number of variable process parameters (time of storing, pressure test, evacuation, etc.). Besides, for the leaking dischargers exposed to atmosphere with the leak related to the molecular flow mode it is necessary to consider atmospheric neon inside the controlled sealed envelope since it affects essentially the calculation-based assessment of sensitivity of the leak tests performed using UFKG.

References

[1] Bushin S A 2011 *Proceedings of 18-th Scientific Workshop «Vacuum Engineering Science» (Sudak)*
[2] Levina L E 1976 *Generalized Empirical Equation for Gas Flow through Leak Channels: Report No. 2408* (Moscow: VNIIVT)
[3] Levina L E 1979 *NDT* **6** 94–8
[4] Bushin S A 2016 *Proceedings of 11th Scientific Workshop «Vacuum Engineering Science» (Moscow)* pp 26–8
[5] Bushin S A 2015 *Vacuum Engineering Science* **2** 71–6
[6] Levina L E and Khavkin L P 1980 *Electronics, Series 7* **1** 108–19
[7] Bushin S A and Galkin S S 2014 *Vacuum Engineering Science* **1** 39–41

Journal of Physics: Conference Series **729** (2016) 012003 doi:10.1088/1742-6596/729/1/012003

Mathematical modeling of the temperature distribution under the cathode spot of the vacuum arc

V G Kuznetsov and E S Babushkina

Institute for problems in mechanical engineering RAS, 61 V. O., Bolshoy pr., Saint-Petersburg, 199178, Russia

E-mail: kvgipme@gmail.com

Abstract. We present a solution to the problem of the temperature distribution under the cathode spot of taking into account melting and spare deposits of metal, brought to boiling temperature on the surface of the cathode spot. The process of heat transfer in the metal is described by the unsteady three dimensional heat conduction equation in Cartesian coordinate system. Similarly, we present a solution to the problem of the temperature distribution in the presence of the pores in the surface layer of the metal. To solve this task we used a numerical method to finite differences and variable directions. We present the calculated data on the distribution of temperature under the cathode spot for different values of spot diameters and speeds its movement.

1. Introduction

Vacuum-arc treatment of surface of materials or surface treatment of the cathode spots of vacuum arcs – the direction of scientific research and practical applications, which in recent years actively developing. In addition to the traditional applications of vacuum-arc treatment to clean the surface of metals and alloys from scale, rust, oil films, technological sediments and other contaminants with the purpose of flaw detection in metal or for the subsequent application of coatings, and new applications. Hardening of the surface layer of metals, the change of physic-chemical characteristics of the surface layer of structural materials, restoration of the plastically deformed surface shape of products, cleaning of surfaces prior to their welding to obtain qualitative weld, the creation of a given surface forms, for example, to retain oil films on the surface, creating on the surface passivation films for corrosion protection – is not a complete list of modern technology using vacuum-arc surface treatment [1–4].

One of the main parameters determining the efficiency of technological processes to the surface by the cathode spot of vacuum-arc discharge is the temperature near the cathode spot and the average temperature of the surface during its movement.

2. Mathematical modeling of temperature distribution

When solving the problem about the temperature distribution under the cathode spot of taking into account melting and evaporation of metal, brought to boiling temperature on the surface of the cathode spot, take into account that the movement of the cathode spot is a intermittent nature. Some time, the spot remains in one place and then jumps to the new plot surface, separated from the initial one or two radius of the spot. Numerically was solved the problem of heat conduction in Cartesian coordinate system for the computational domain presented in figure 1.

The process of heat transfer in the metal is described by the unsteady three-dimensional heat conduction equation in Cartesian coordinate system as:

$$\rho c_p \frac{\partial T}{\partial t} = \lambda \left(\frac{\partial^2 T}{\partial x^2} + \frac{\partial^2 T}{\partial y^2} + \frac{\partial^2 T}{\partial z^2} \right), \tag{1}$$

where ρ – is the material density; λ – thermal conductivity coefficient; c_p – is the heat capacity; T – is the temperature; x, y, z – coordinates.

Since the task has a plane of symmetry, the solution was carried out in half the expected volume. The cathode spot is moved along the plane B (figure 1). The plane A was chosen as the plane of symmetry and on it was set the boundary conditions of symmetry, i. e. no flow through the surface:

$$\lambda \left. \frac{\partial T}{\partial s} \right|_A = 0. \tag{2}$$

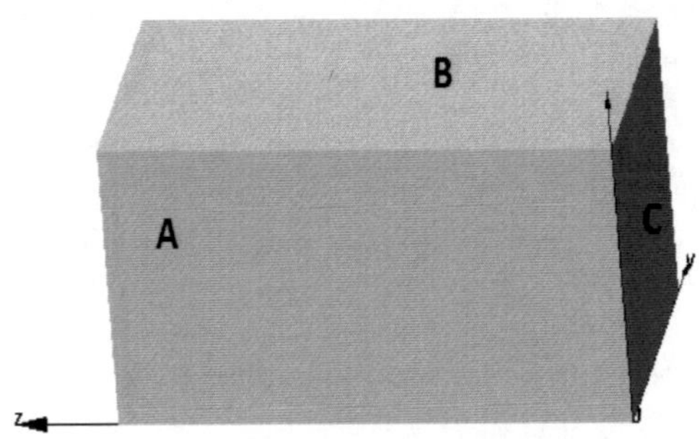

Figure 1. An example of the calculated region.

The surface B faces towards the vacuum and it was set to a boundary condition of radiation:

$$\lambda \left. \frac{\partial T}{\partial s} \right|_B = \sigma \varepsilon \left(T_\infty^4 - T^4 \right), \tag{3}$$

where T_∞ – ambient temperature; σ – Stefan–Boltzmann constant; ε – is the integral coefficient of radiation.

The effect of cathode spots was recorded as heat flow in the metal surface on the surface area B. The heat flow is proportional to the voltage U on the vacuum arc, to the current intensity I and to the power cathode coefficient h_k, then:

$$\lambda \left. \frac{\partial T}{\partial s} \right|_S = \sigma \varepsilon \left(T_\infty^4 - T^4 \right) + \frac{UIh_k}{S}, \tag{4}$$

where S – is the area of the cathode spot.

Considering that the size of the computational domain much larger than the size of cathode spots and the surface C (figure 1) sufficiently far from the heat source, then $T|_C = T_0$, where T_0 – the initial temperature of the metal.

When solving the problem was taken into account melting and evaporation of metals under the action of a cathode spot. To account of these phenomena consider the phase transition from the solid to the liquid state. The transition from liquid to gaseous phase is carried out similarly. The transition into the liquid phase is carried out when the temperature of metal the melting temperature T_{pl}. In addition, for the transition it is necessary to spend certain amount of thermal energy Q_{pl}. The value of the specific heat of melting is defined as $\lambda_{pl} = Q_{pl}/m$, where m – is the mass of the consumable substance.

Journal of Physics: Conference Series **729** (2016) 012003 doi:10.1088/1742-6596/729/1/012003

The solution of equations was performed numerically. The computational domain was divided differential three-dimensional grid in the directions *x*, *y*, *z*. To solve hocks set tasks were used the numerical method of finite differences and variable directions.

Figure 2 presents a picture of the thermal field surrounding the crater of the cathode spot, at a current arc I = 100 A, the cathode spot radius $r = 5\cdot10^{-5}$ m, the speed of movement of the cathode spots of V = 1 m/s.

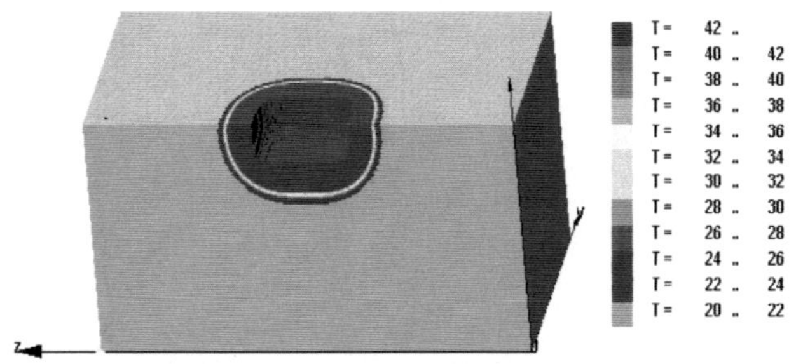

Figure 2. Example of calculation of a picture of the thermal field cathode spot.

In numerical experiments we have investigated the temperature distribution under the cathode spot at the moment when the cathode spot moves to a different place. This point corresponds to the maximum temperature under the cathode spot, which is approximately equal to the boiling temperature of the metal. As variable parameters were taken the radius of the cathode spot, the velocity of the spot and the current of the vacuum arc.

Table 1 presents the calculated values of residence time of the cathode spot on one place on its radius and speed.

Table 1. The results of the calculation of residence time of the cathode spot, s.

V, m/s		r, m	
	$5\cdot10^{-5}$	$1\cdot10^{-5}$	$3\cdot10^{-6}$
1	$5\cdot10^{-5}$	$1\cdot10^{-5}$	$3\cdot10^{-6}$
10	$2{,}5\cdot10^{-6}$	$5\cdot10^{7}$	$1{,}5\cdot10^{-7}$
100	$1\cdot10^{-6}$	$2\cdot10^{-7}$	$6\cdot10^{-8}$

The calculated temperature distribution from the surface under the cathode spot is presented in figure 3. Surface in this case is considered to be the deepest point of the molten hole.

The cathode spot of vacuum-arc discharge moving along the surface of the cathode, it affects how a localized, highly concentrated source of energy. The temperature of the material at the surface in the spot area, as a rule, exceeds the temperature boil. If the speed of tens meters per second the heat affected zone in the metal around the cathode spot is small. Calculations show that the temperature distribution from the surface under the cathode spot is plummeting conductive and at a distance equal to the radius of the spot, the temperature approaches to the temperature of the bulk metal. In the presence on the cathode surface of thick oxide films in the form of, for example, scale or rust, when the cathode spot is used to clean these films, the speed of movement of the spots decreases significantly and may reach units of centimeters per second. The zone of thermal action on the metal is greatly increased. In the presence of near surface layers of metal cavities in the form of pores, cracks, etc. heat sink from the zone of the cathode spot to a mass of metal is deteriorating, the cathode spot can thin melted surface wall and open surface cavity. Such situations often occur in a vacuum-arc cleaning the surface of various products, for example, if a poor quality casting (figure 4).

This figure represents a photograph of the interior surface of the housing axle box truck of a railroad car, which are made with the help of technology of the “ground” cast. Cleaning was carried out on the installation in which the axle box housing, having a cylinder form, which acted as vacuum chamber. Vacuum arc cleaning has been used as an effective method of flaw detection of the metal.

Consider the amount of metal with sides A, B, C, within which cavity are D (figure 5). The cathode spot moves on the surface B, which is in a vacuum.

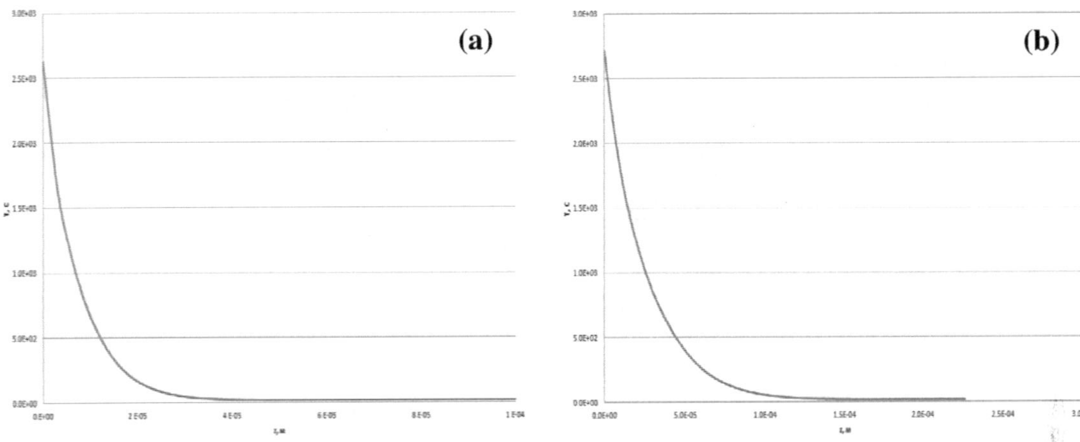

Figure 3. The temperature vs. distance from the surface under the cathode spot in different speeds of motion of cathode spots (I = 300 A, $r = 10^{-4}$ m): (a) – V = 10 m/s; (b) – V = 1 m/s.

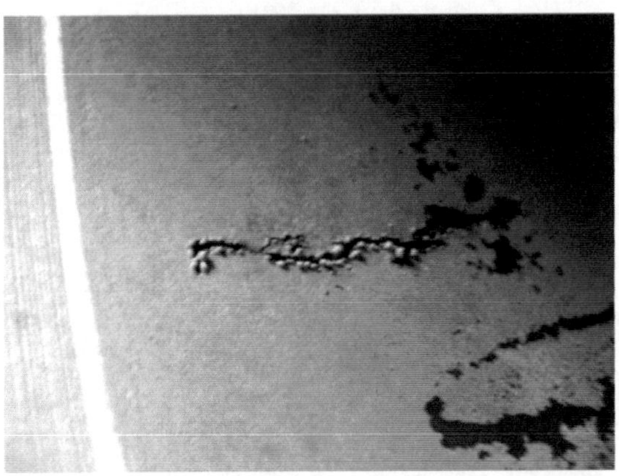

Figure 4. Uncovered near the surface the pore as a result of vacuum-arc cleaning of axle box housing for railway car.

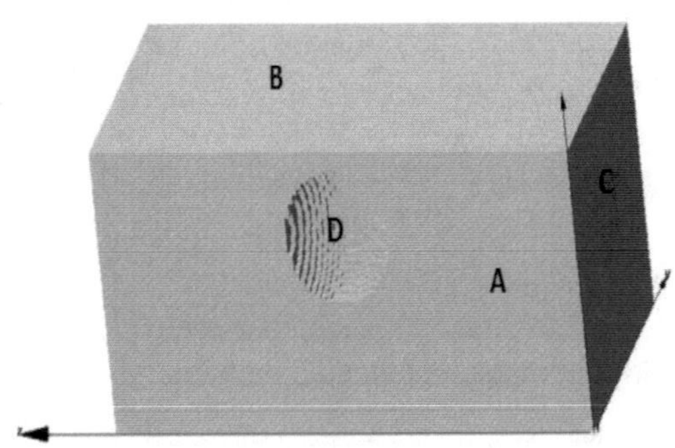

Figure 5. An example of the calculated region with a pore.

For the computational domain (figure 5) was numerically solved the problem of heat conduction. The power fed from the discharge to the cathode (the surface) is proportionality to current of vacuum arc I, the voltage drop across the discharge U and the coefficient of cathode power h_k. The process of heat transfer in the metal is described by three-dimensional non-stationary heat conduction equation in Cartesian coordinate system. From surface power is given also by radiation to the side of the vacuum volume. On the surface of the cavity D was defined the condition of radiation:

$$\lambda \left. \frac{\partial T}{\partial s} \right|_{D} = \sigma\varepsilon\left(T_{av}^4 - T^4\right), \quad (5)$$

where T_{av} – the average temperature of the cavity surface.

Were taken into account melting and evaporation of metals under the action of a cathode spot. The solution of equations was performed numerically.

Was studied, in that the distance h a cavity inside the metal may be situated, so the cathode spot could have melted through the metal, separated the surface from the cavity, during its stay on one place. Figure 6 shows an example of calculation of fields of temperature and depth of melting of the metal under the cathode spot.

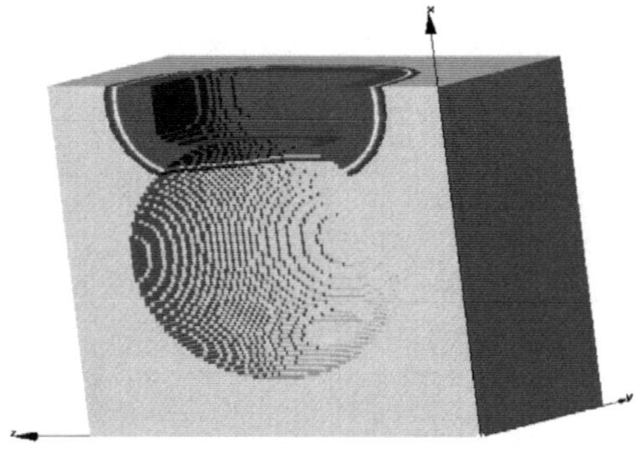

Figure 6. The example of calculations of temperature fields and the depth of penetration of metal under the cathode spot to the cavity: $V = 1$ m/s, $r = 5 \cdot 10^{-5}$ m.

Table 2 presents the results of calculations of maximum depth of metal, which the cathode spot is melts, while over at cavity, from speed of movement of the spot and its radius.

Table 2. The results of calculations of maximum depth of metal, which the cathode spot is melts, m.

V, m/s	r, m		
	$5\cdot10^{-5}$	$1\cdot10^{-5}$	$3\cdot10^{-6}$
1	$1\cdot10^{-4}$	$7\cdot10^{-5}$	$4,5\cdot10^{-5}$
20	$3,2\cdot10^{-5}$	$2,4\cdot10^{-5}$	$1,7\cdot10^{-5}$
50	$2\cdot10^{-5}$	$1,8\cdot10^{-5}$	$1,3\cdot10^{-5}$

3. Conclusions

The results of the calculations were used to assess the efficiency of some technological processes in the design of the equipment.

References

[1] Steshenkova N A, Shumilov V P, Kuznetsov V G, Shalimov A G and Sokolov O G 2005 *Bulletin of shipbuilding technology* **13** 43–50

[2] Atroshenko S A, Kuznetsov V G and Salakhova N R 2012 *Hardening technology and coatings* **11** 3–11

[3] Babushkina E S, Bessonov N M and Kuznetsov V G 2012 *Vacuum technique and technology* **2** 93–4

[4] Kuznetsov V G, Kurbanov T A, Krymov D V and Ponimatkin V P 2014 *Vacuum technique and technology* **1** 78–81

Journal of Physics: Conference Series **729** (2016) 012004 doi:10.1088/1742-6596/729/1/012004

Characteristics of a glow discharge maintained in the vapors of a liquid

A N Ramazanov[1], D K Kostrin[1], V D Goncharov[2] and A A Lisenkov[3]

[1] Department of electronic instruments and devices, Saint Petersburg Electrotechnical University "LETI", 197376, Saint Petersburg, Russia
[2] Department of theoretical fundamentals of electrical engineering, Saint Petersburg Electrotechnical University "LETI", 197376, Saint Petersburg, Russia
[3] Institute of Problems of Mechanical Engineering, Russian Academy of Sciences, 199178, Saint Petersburg, Russia

E-mail: ramazanow.alexander@yandex.ru

Abstract. Glow discharge in atmospheric pressure air in the discharge cell with a liquid cathode is considered. Current-voltage characteristic of such a system at different distances between the electrodes and various electrode placements, different thickness of the liquid layer and the cathode surface area are shown. Spectral emission of the glow discharge plasma in the vapors of a liquid is demonstrated.

In this work a system composed of the anode and submerged under the liquid layer cathode was investigated (figure 1). For ignition of the discharge was developed a high voltage scheme, which allows to vary the pulse repetition rate from 15 to 50 kHz, at a voltage from 0 to 3,5 kV (figure 2). Rod of tungsten with a diameter of 4 mm, sharpened at an angle of 30°, separated from the liquid by an air gap in which discharge occurs upon application of high voltage, is used as the anode.

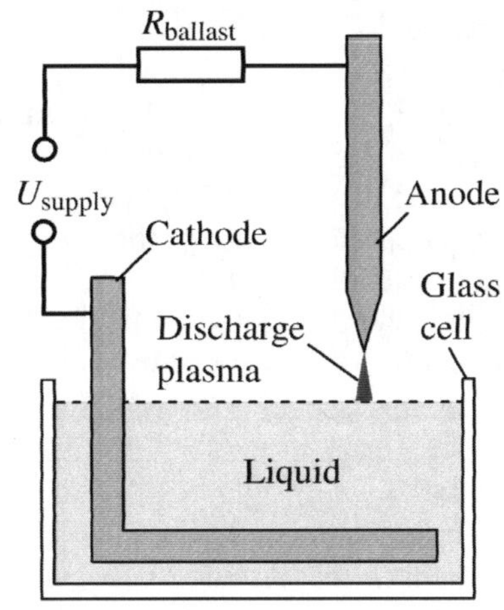

Figure 1. Simplified design scheme of the discharge cell.

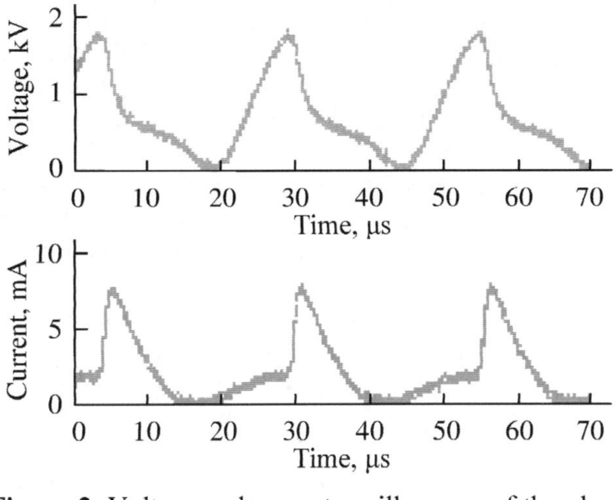

Figure 2. Voltage and current oscillograms of the glow discharge.

While using a similar supply system conductive liquid plays the role of distributed resistive-capacitive ballast. Distance from the anode to the surface of the liquid was varied in the range 3...10 mm. In the case where it is not specified the thickness of the liquid layer above the metal cathode in the experiments was equal to 4 cm.

Current-voltage characteristics (I–V curves) of the studied system (the glow discharge and the distributed liquid ballast) at different distances *d* between the anode and the surface of the liquid are shown in figure 3. The results were obtained at a constant frequency of 40 kHz. Similar results were observed in the whole investigated frequency range.

At d = 4...6 mm the discharge was burning only in the certain ranges of frequencies and voltages. At d = 5 mm frequency range of the discharge was 19,5...50 kHz. At d = 3 mm the discharge was maintained over the entire range of frequencies and voltages. The high brightness of the glow discharge is observed at currents exceeding 20 mA.

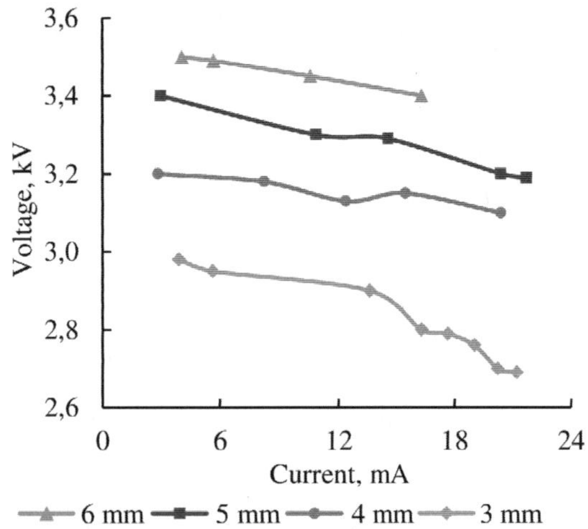

Figure 3. I–V curves of the glow discharge at different distances metal anode–liquid cathode.

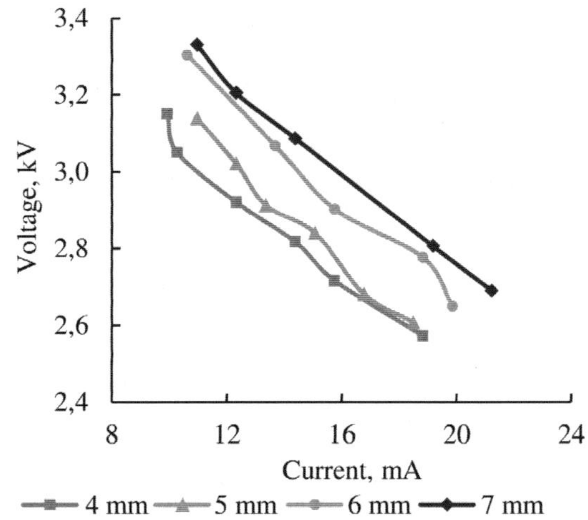

Figure 4. I–V curves of the glow discharge at different thickness of the liquid layer above the metal cathode.

As the distance between the liquid surface and the cathode is changed I–V curves of the discharge become more steeply dipping (figure 4). As a result the energy contributed into the discharge decreases and, consequently, the brightness of its radiance reduces. In this experiment the distance between the metal anode and the liquid was maintained equal to 3 mm.

The increase in the current flowing in the discharge gap leads to a change in the shape of the discharge – from cone, in which there is a uniform current distribution on the surface of the liquid, to ellipse, in which there is a "binding" of the discharge to a specific area on the surface of the liquid. In this case the distributed resistance of the liquid layer is not enough to limit the current and instabilities, inherent to the glow discharge, lead to its contraction and the formation on the surface of the liquid of an explicit cathode spot [1]. This spot move quickly across the surface of the liquid.

In the next experiment the influence of the surface area of the metal cathode located in the depth of the liquid on the I–V curves of the gas discharge was investigated (figure 5). Originally the area of a rectangular plate was equal to about 14 mm^2. With the decrease in the surface area of the metal cathode decreases the voltage of the discharge sustaining, this fact is connected with the decrease of the resistance of the liquid volume through which an electric current flows.

Similar results can be obtained while shifting the anode relative to the cathode. The results of this experiment are shown in figure 6. With the shift of the anode increases the distance between the electrodes, which leads to an increase of the discharge voltage, and also increases the distributed resistance of the liquid, which leads to a sharper voltage drop with the increasing current.

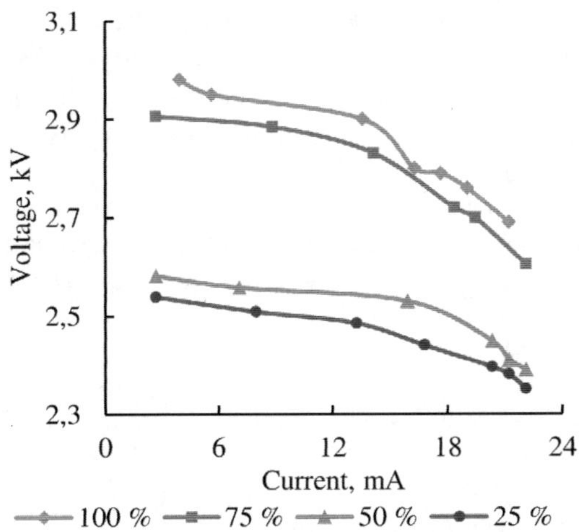

Figure 5. I–V curves of the glow discharge at different surface areas of the metal cathode.

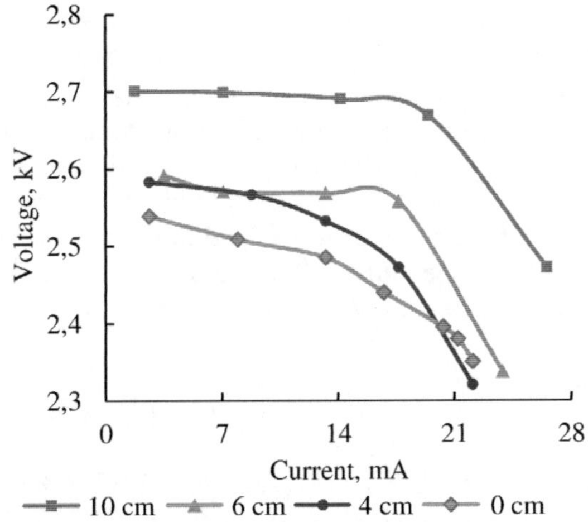

Figure 6. I–V curves of the glow discharge at different positions of the anode relative to the metal cathode.

Spectrogram of the atmospheric pressure discharge in the air between the metal anode and the liquid surface was received using a spectrometer ISM3600 [2, 3]. Processing of the obtained spectra was performed using the Aspect2010 software [4]. One of the main functions of this software is plasma emission spectrum analysis. As the test liquid in all the experiments tap water was used.

In figure 7 a spectrogram of the glow discharge with liquid cathode at a distance between the anode and the water surface d = 4 mm and the frequency of the supply voltage of 30 kHz is shown. The sensitivity of the spectrometer (charge accumulation time) is equal to 2 s.

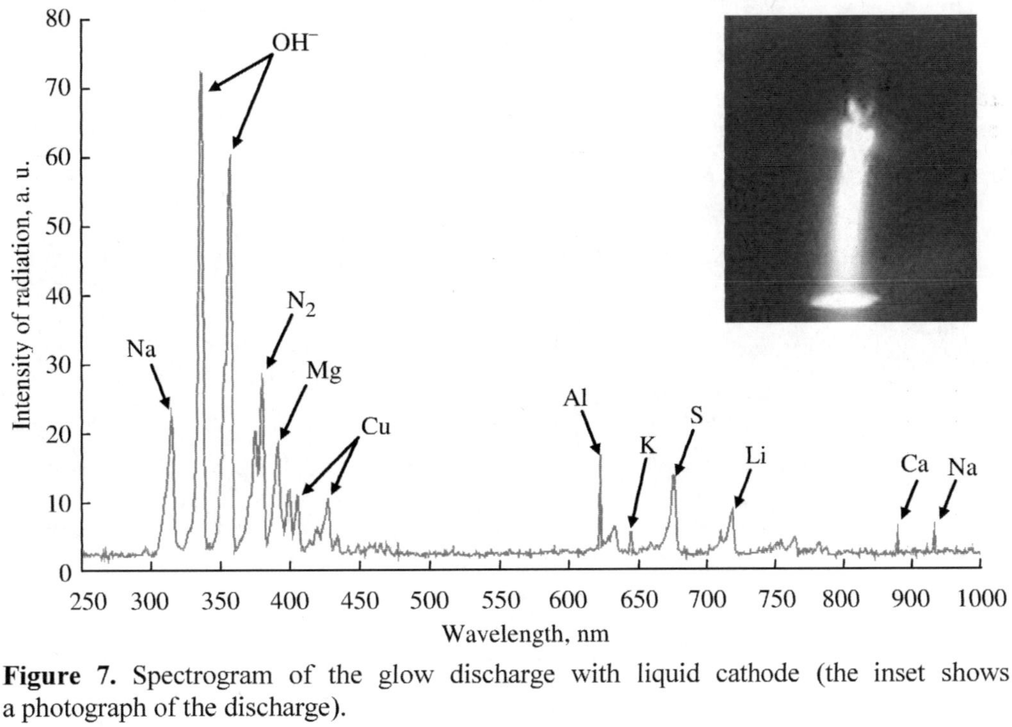

Figure 7. Spectrogram of the glow discharge with liquid cathode (the inset shows a photograph of the discharge).

The most intense emission spectral lines of the glow discharge plasma are concentrated in the interval of wavelengths 300...400 nm and correspond mainly to the hydroxyl group OH^-, singly charged ions of nitrogen N^+ and oxygen O^+ and also molecular nitrogen and oxygen. In the visible wavelength range spectral lines of molecules of lithium, copper, potassium, sulphur and aluminum are observed. In the range 800...1000 nm peaks corresponding to calcium and sodium can be seen.

The efficiency of the spectral lines excitation for various elements is different and strongly depends on the electric parameters of the discharge and the structural characteristics of the discharge cell and even the temperature of the liquid. During the experiment it was found that the use of more hot water shifts the I–V curves of the discharge in the direction of lower voltages.

References

[1] Bystrov Yu A, Kostrin D K, Lisenkov A A and Vetrov N Z 2015 *Vakuum in Forschung und Praxis* **2** 22–5
[2] Kostrin D K and Oukhov A A 2013 *Biotechnosphere* **3** 21–5
[3] Kostrin D K, Uhov A A and Lisenkov A A 2016 *Vakuum in Forschung und Praxis* **3** 34–7
[4] Uhov A A, Gerasimov V A, Kostrin D K and Selivanov L M 2014 *Journal of Physics: Conference Series* **567** 012039

Characteristics of flows of energetic atoms reflected from metal targets during ion bombardment

A Kuzmichev[1], V Perevertaylo[1], L Tsybulsky[1] and O Volpian[2]

[1] National Technical University of Ukraine "Kiev Polytechnic Institute", 37 prospect Pobedy, Kiev 03056, Ukraine
[2] Scientific-Manufacturing Enterprise "Fotron-Auto Ltd.", 8 Novodanilovskaya naberezhnaya, Moscow 117105, Russia

E-mail: A.Kuzmichev@edd.ntu-kpi.kiev.ua

Abstract. Particle number and energy reflection coefficients for energetic neutralized gas ions (Ar and O atoms) backscattered from metal targets during ion bombardment have been calculated using TRIM code. The energy distributions of reflected atoms are computed, too, and their dependence on the primary ion energy and the angle of ion incidence is determined. The obtained data confirm the possibility of employing energetic atoms reflection for generation of high energy neutral beams and point out to take this phenomenon into account under analysis of the ion technology for coating deposition.

1. Introduction

Gas ion bombardment of metal targets is widely used in electronic and optical industry for sputtering of metal atoms and deposition of coating onto substrates. This process is accompanied by partial reflection/backscattering of bombarding ions from the target surface [1, 2]. The reflected particles practically all are electrically neutral since the primary projectile ions are neutralized during the contact with metal surface. Besides, the primary molecular ions dissociate; so, the reflected particles are gas atoms and they may have significant part of the primary ion energy. These energetic atoms bombard the substrates, heat them and affect the microstructure of the coatings [3]. The reflection and neutralization of ions on metal surface may be used for generation of directed high-energy neutral atom beams [4–6]. Such beams are employed for sputtering of dielectrics and semiconductors [4–6].

The phenomenon of reflection/backscattering of energetic atoms from metal surface during ion bombardment was considered in literature and several computer codes have been developed for simulating this process [1, 2, 7]. It has been shown that the reflection coefficients may achieve tens of percents, especially at oblique incidence. The published information relates mainly to reflection of light (H and He) atoms [2, 7]. However from practical point of view the most interesting for sputtering technology is data on reflection of such working particles as argon and oxygen ions. The task of this work is to calculate the characteristics of energetic flows of argon and oxygen atoms reflected from different metal targets during bombardment by ions with energy in range of 0,5–5,0 keV used for sputtering. The set of light and heavy metals (from Si, Al and Ti to Ta) was chosen as the typical target set used in sputtering technology.

2. Calculation of energetic atom flows characteristics and discussion

The calculations were performed using the TRIM code (TRansport of Ions in Matter) [7]. It is based on the Monte Carlo method to follow a random walks of N_0 = 99 500 high-energy particles (the former projectile ions) in a solid target using the approximation of binary collisions of the walking particles with the target atoms. The trajectory of each projectile particle is traced as long as it does not lose its energy. Particle number and energy reflection coefficients (respectively, R_N and R_E) are used as integrated characteristics of the process. The coefficient R_N is defined as the ratio of reflected/backscattered N atoms to the number N_0 of primary atom particles bombarding the target. In the case of one-atom ions (Ar^+ or O^+) the value N_0 equals the quantity of ions. In the case of two-atom molecular ions (O^+_2) we must take N_0 as double quantity of ions because these ions dissociate and then two atoms walk within the target body as independent particles. The presence of charge at the bombarding particles does not affect R_N [2]. Table 1 and table 2 show the results of R_N and R_E calculations, where E_0 is the initial energy of the bombarding particles in terms of atoms.

Table 1. Calculation of the particle number and energy reflection coefficients for the oxygen ions (O^+) at normal incidence.

E_0, keV	Reflection coefficients, %	Material and atomic mass of the target					
		Al 26,98	Ti 47,90	Zn 65,38	Zr 91,22	Nb 92,91	Ta 180,95
0,5	R_N	5,02	12,82	20,65	22,17	23,53	36,53
	R_E	0,41	2,14	5,18	6,10	6,65	15,55
1,0	R_N	3,98	11,43	18,85	20,15	21,24	35,45
	R_E	0,33	1,99	4,74	5,58	5,98	14,81
5,0	R_N	2,44	7,74	14,36	16,22	16,50	31,06
	R_E	0,21	1,33	3,65	4,35	4,55	12,82

Table 2. Calculation of the particle number and energy reflection coefficients for the argon ions (Ar^+) at normal incidence.

E_0, keV	Reflection coefficients, %	Material and atomic mass of the target					
		Al 26,98	Ti 47,90	Zn 65,38	Zr 91,22	Nb 92,91	Ta 180,95
0,5	R_N	0,0834	2,90	9,25	12,73	13,53	28,40
	R_E	0,0014	0,12	0,74	1,57	1,79	7,98
1,0	R_N	0,0724	2,66	7,91	1,52	12,41	26,87
	R_E	0,0010	0,11	0,60	1,45	1,65	7,28
5,0	R_N	0,0673	1,75	5,07	8,77	9,25	22,08
	R_E	0,0010	0,08	0,44	1,12	1,23	6,15

As one can see, the R_N is from several percent to tens percent (subject to the target material), which is a very large value. The lighter bombarding ions, the less energy E_0 and the heavier target atoms, the bigger R_N is (see coefficients for Al, Ti and Ta). The backscattering from the silicon target was very

Journal of Physics: Conference Series **729** (2016) 012005 doi:10.1088/1742-6596/729/1/012005

small and it is not presented in the tables. Energy reflection coefficient R_E also lies in the range from several percent to tens percent, but it is several times smaller than the R_N.

The increase of the ion incidence angle φ leads to significant increase of R_N (see figure 1) due to less penetration depth of the projectile particle into the target body at oblique incidence and, hence, greater chance for the particle to return back to the target surface and to go out from the target. The less the way of the particle inside of the target, the less the loss of particle energy and the higher the energy of the backscattered atom are. Thus, the reflected atoms have greater energies at oblique ion incidence than at normal incidence (φ = 0°). The latter is clearly demonstrated by figure 2 and figure 3, which depict the energy distributions of oxygen atoms, which are reflected/backscattered from the targets of titanium and tantalum.

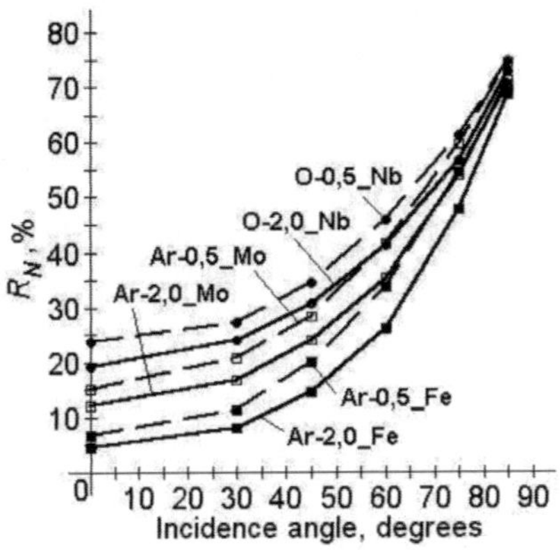

Figure 1. Dependences of particle number reflection coefficient R_N on incidence angle φ for O^+ and Ar^+ ions with initial energy E_0 of 0.5 keV and 2 keV. Target materials are Fe, Mo and Nb.

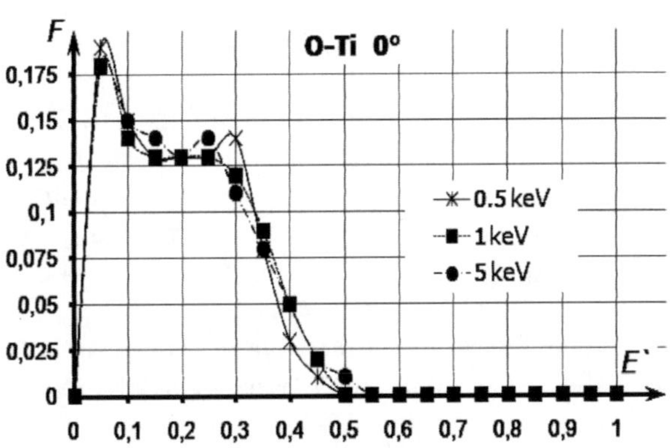

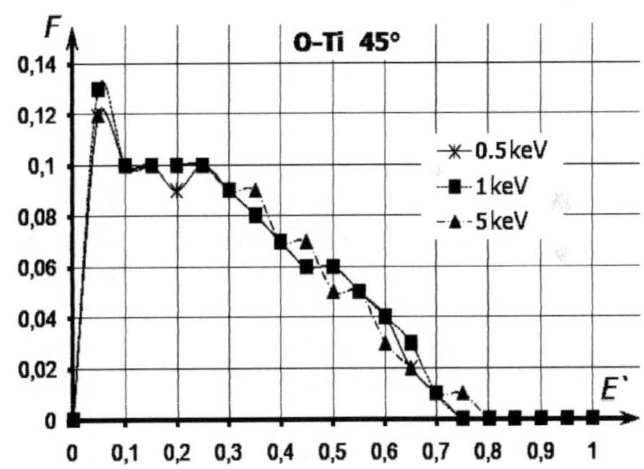

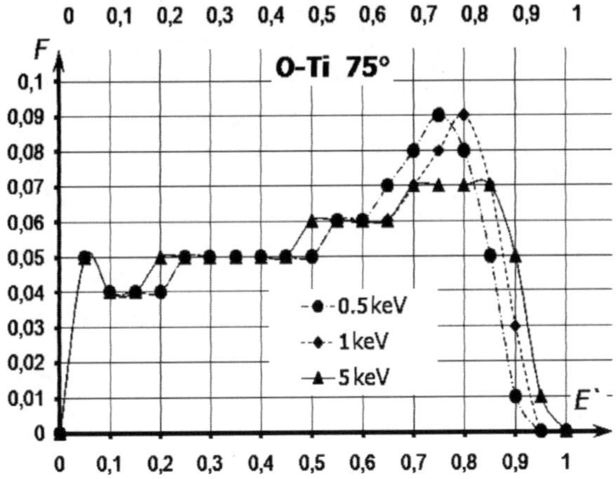

Figure 2. Energy distributions of oxygen atoms reflected from the titanium target at different incidence angles. $E' = E/E_0$.

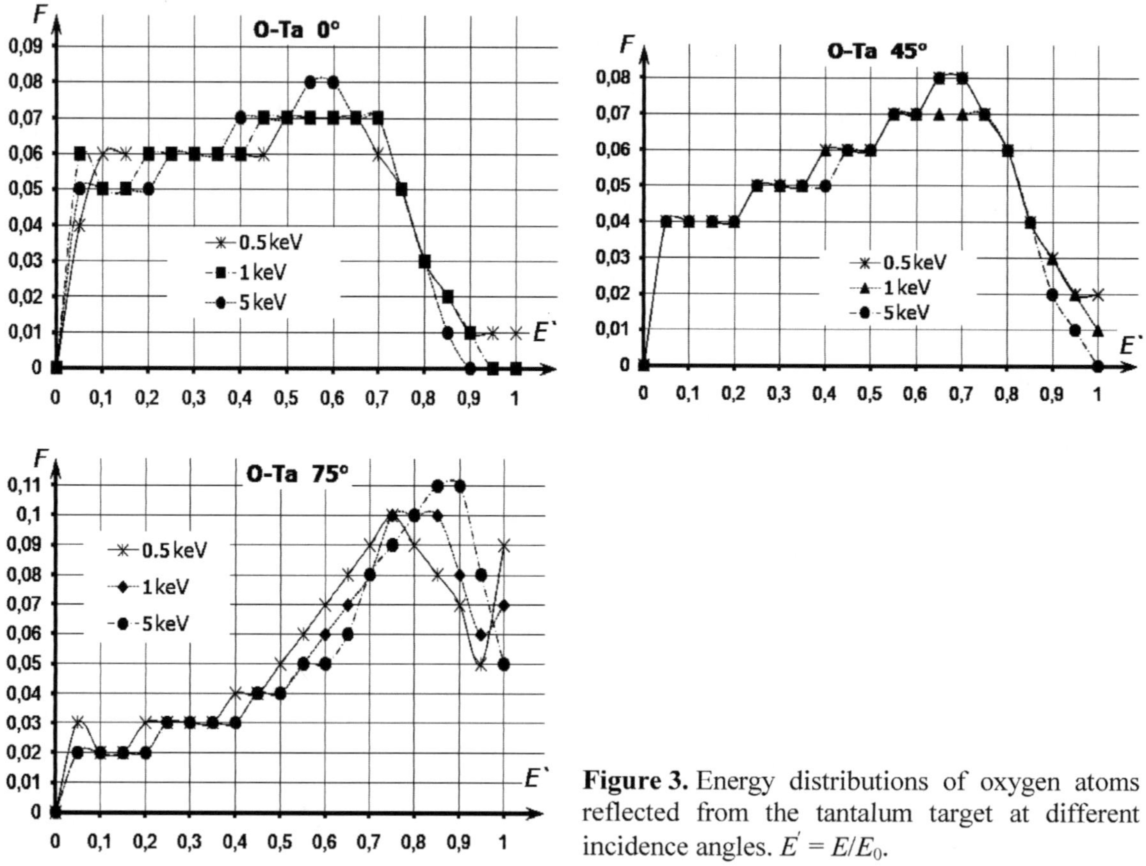

Figure 3. Energy distributions of oxygen atoms reflected from the tantalum target at different incidence angles. $E^{'} = E/E_0$.

Figure 2 and figure 3 present the results of the calculation of energy distributions of the reflected atoms for three values of the bombarding ion energy (0,5, 1 and 5 keV) and three values of the incidence angle (0, 45 and 75 degrees). Results are presented as a polygon. Here, $E^{'} = E/E_0$ is relative energy of the reflected atoms and F was defined as the ratio of number of atoms having an energy in the range of $[E^{'} - 0{,}05,\ E^{'}]$ to the total amount of the reflected atoms N. The entire range of possible energies was divided into 20 parts, so that the sum of 20 discrete values of F is equal to 1.

As one can see, the general form of distributions of reflected atoms on the relative energy is not strongly dependent on the value of the initial energy E_0 of the bombarding ions in the range of 0,5–5,0 keV. However, the increase of incidence angle φ strongly influences on the energy distributions. If $\varphi \sim 0°$ the distributions have a squared shape, when $\varphi = 75°$ their shape becomes triangular-like with a large number of atoms with $E' \sim 0{,}7–1{,}0$. Atoms flows, which are reflected from the target with big atomic mass, contain many high-energy particles. This is particularly valuable for producing beams of energetic neutral atoms from ion beams by ion neutralization of on the metal surface [4–6].

3. Conclusions

Thus, the results of calculations show the significant flows of energetic Ar and O atoms reflected/backscattered from the metal target are generated in the sputtering process with participation of both the heavy argon ions and the lighter oxygen ions. Energy of reflected atoms, which is transferred to the substrate surface, may significantly affect the formation and characteristics of the deposited coatings; in particular, it may strongly affect the refractive index profile during deposition of gradient optical coatings [8]. So, it is necessary to take into account the reflection of energetic atoms and transfer of their energy to substrates under analysis of coating deposition conditions. The obtained data on the reflection of energetic gas neutralized particles can also be used during design of sources of neutral atom beams with neutralizing the primary ion beams on the metal surface. The use of heavy metals is more preferable for building the neutralizers.

Acknowledgments
This work was financially supported by the grant of the Ministry of Education and Science of the Russian Federation. (Scientific project "Creating electro-optical gradient thin-film structures for precision optics and analytical instrumentation". Agreement № 14.579.21.0066 from 23.10.2014 for the grant with the Ministry of Education and Science of Russian Federation, the unique identifier RFMEFI 57914 X 0066).

References

[1] Ziegler J F, Biersack J P and Littmark U 1985 *The Stopping and Range of Ions in Matter* (New York: Pergamon)
[2] Kurnaev V A, Mashkova E S and Molchanov V A 1985 *Light Ion Reflection from Solid Surface* (Moscow: Energoatomizdat)
[3] Sarakinos K, Alami J, Karimi P M, Severin D and Wuttig M 2007 *J. Phys. D Appl. Phys.* **40** 778–85
[4] Barchenko V T, Lisenkov A A and Babinov N A 2014 *J Phys Conf. Ser.* **567** 012029
[5] Kim S J, Wang S J, Lee J K, Lee D H and Yeom G Y 2004 *J. Vac. Sci. Technol.* A **22** 1948–55
[6] Park B J *et al* 2008 *J. Phys. D Appl. Phys.* **41** 024005
[7] Eckstein W 1991 *Computer Simulation of Ion-Solid Interactions* (Berlin: Springer-Verlag)
[8] Volpian O D, Kuzmichev A I and Obod Yu A 2015 *Inorganic Materials: Applied Research* **6** 234–9

The target heating influence on the reactive magnetron sputtering process

A Bondarenko, A Kolomiytsev and V Shapovalov

St. Petersburg State Electrotechnical University "LETI", 5 Prof. Popov Street, St. Petersburg 194354, Russia

E-mail: stopnastia@gmail.com

Abstract. A physicochemical model for the reactive magnetron sputtering of a "hot" target is described in this paper. The system consisting of eight algebraic equations was solved for a tantalum target sputtered in an O_2 environment. It was established that the hysteresis effect disappears with the increase of the ion current density.

1. Introduction

The reactive magnetron sputtering technology is one of the most commonly used methods of the deposition of thin films [1]. By changing the reactive gas flow rate and the current density on the target, it is possible to deposit both metallic and stoichiometric thin films. Using the magnetron with a "hot" target opens new possibilities for the reactive sputtering, since the temperature of the target can be led up to its melting point [2, 3].

For the effective application of the reactive sputtering technology, it is necessary to establish its general characteristics and to figure out the relations between the dependent and independent variables. This research was carried out both experimentally and by simulations. All notorious models for the reactive sputtering technology do not consider the target heating effect and are based on two basic assumptions [4–6]:

- two processes occur on the target surface: the formation of a thin layer of the metal/reactive gas compound and the sputtering of this compound by the argon ions;
- the deposition of both metal and compound particles sputtered from the target and the chemical reaction occur on the substrate and the walls of the vacuum chamber.

2. Model

The physicochemical model offered in [5, 6] describes this process in the most correct way. We are going to use this model considering the effect of the heating of the target surface.

The metallic (M) target is sputtered in an Ar + X_2 environment, where X_2 is a chemically active gas. The stationary state of the process is provided by nonisothermal conditions, since the substrate and the walls of the vacuum chamber have constant temperatures T_s and T_w, respectfully. The target temperature T_t depends on the power density at the target surface (or the ion current density j_+). Furthermore

- the discharge current density represents a summation

$$j = (1+\gamma)j_+ + j_T, \tag{1}$$

where γ is the ion-induced electron emission coefficient; j_T is the thermal electron emission current density (Richardson's law);

- the M_mX_n compound appears on all of the surfaces as a result of a chemical reaction:

$$\mathrm{M}+\frac{n}{2m}\mathrm{X}_2 \overset{k(T)}{\leftrightarrow} \frac{1}{m}\mathrm{M}_m\mathrm{X}_n, \tag{2}$$

where $k(T)$ is the reaction rate coefficient (Arrhenius equation). The reaction (2) happens due to the physical adsorption of the X_2 molecules, which is described by the Langmuir isotherm. As a result, the consumption of the X_2 gas occurring on the target, substrate and walls surfaces is denoted as Q_t, Q_s and Q_w, respectively;

- the target surface is sputtered by the argon ions and evaporated according to the Hertz–Knudsen law, hence the metal F_{1t} and the compound F_{2t} flux rates can be expressed as

$$F_{1t}(j_+)=F_{1sp}(j_+)+F_{1ev}\left[T_t(j_+)\right]; \tag{3}$$

$$F_{2t}(j_+)=F_{2sp}(j_+)+F_{2ev}\left[T_t(j_+)\right]. \tag{4}$$

The first addends in the right part of equations (3) and (4) describe the sputtered particles, the second ones describe the evaporated particles;

- the temperature influence on such parameters of the process as the sputter yield and the ion-induced electron emission coefficient is neglected.

Using the law of mass action the way it was done in [5] and taking into consideration (1)–(4) we obtained a system of eight algebraic equations for one reactive gas. These equations describe

1) the balance on the surfaces of

- the target:

$$F_{ch}\left[T_t(j_+)\right]+F_{2t}(j_+)=0; \tag{5}$$

- the substrate and the walls:

$$F_{ch}(T_i)+F_{2i}(j_+)-F_{1i}(j_+)=0,\ i=\mathrm{s, w}; \tag{6}$$

2) the reactive gas flow on all of the three internal surfaces of the vacuum chamber:

$$Q_i=\frac{n}{2}F_{ch}(T_i)A_i,\ i=\mathrm{t, s, w}; \tag{7}$$

3) the reactive gas evacuated through the pumping system:

$$Q_p=c^0 pS_p; \tag{8}$$

4) the balance of the reactive gas flows:

$$Q_0=Q_t+Q_s+Q_w+Q_p; \tag{9}$$

The following parameters were used in (5)–(9): $F_{ch}(T_i)$ is the temperature dependable flux rate of the M_mX_n compound formed due to the chemical reaction (2) on the i-surface; $F_{1i}(j_+)$ and $F_{2i}(j_+)$ are the flux rates of the metal and the compound particles sputtered from the target and deposited on the i-surface; $c^0 = 2.5\cdot10^{25}$ Pa^{-1}m^{-3} is a coefficient converting the flow Q_p from [Pa · m^3/s] to [s^{-1}]; S_p is the system pumping speed; p is the oxygen partial pressure.

The system of algebraic equations (5)–(9) contains two independent variables: the reactive gas flow and the ion current density. The unknowns are the partial pressure of the reactive gas X_2, the gas flows Q_t, Q_s, Q_w and Q_p and the fraction of each surface covered with the M_mX_n compound. The system has to be solved numerically. In particular, when the surface temperature is so small that $j_T << \gamma j_+$ (1), $F_{1ev} << F_{1sp}$ (3) and $F_{2ev} << F_{2sp}$ (4), this system of equations describes the reactive sputtering process when $T_t \neq f(j_+)$.

3. Simulations

The system of equations (5)–(9) was numerically solved for the magnetron with a tantalum target sputtered in the Ar + O_2 environment. The power density dependence of the target temperature was obtained by solving a thermal task. Using this dependence and the experimental current-voltage characteristics we determined the relation connecting the target temperature and the ion current density:

$$T_t \approx 293 + 1517\left(1 - e^{-0.00438 j_+}\right). \tag{10}$$

Considering (10) the equations (5)–(9) were solved with the ion current density ranging from 0 to 200 A/m^2. Figure 1 shows the oxygen partial pressure vs. the supply of the reactive gas for different ion current densities. Two characteristic curve segments are observed:

- first – when the oxygen flow is small, the pressure is close to 0;
- second – the pressure is proportional to the gas flow.

Similar dependences were often obtained experimentally during the reactive sputtering of a "cold" target [5]. The first segment of the curve is related to the metal mode of the target, the second one stands for the oxide mode. But the main feature of the sputtering process observed during these experiments is the hysteresis effect. The simulations result reflecting this effect is shown in figure 1(a). During the sputtering of a "cold" target the increase of the ion current density does not change the nature of the sputtering process curves. Both instability points of the $p = f(Q_0)$ curve shift to higher Q_0 values proportionally to j_+. The model predicts the elimination of the hysteresis effect at a certain critical value of the ion current density (figure 1(b)).

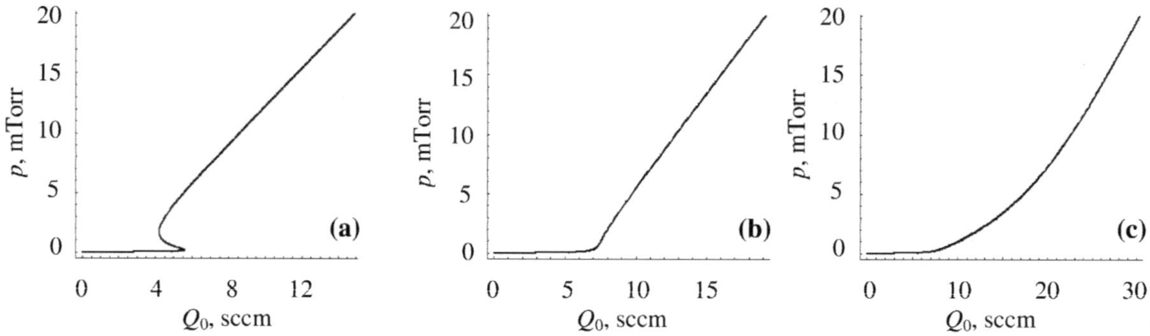

Figure 1. The oxygen pressure vs. the oxygen flow for j_+ (A/m^2): (a) – 160; (b) – 170; (c) – 180.

Acknowledgments
The authors thank Russian Science Foundation for their support on the study (grant 15-19-00076).

References

[1] Levitskii V S, Shapovalov V I, Komlev A E *et al* 2015 *Techn. Phys. Lett.* **41** 1094–6
[2] Lapshin A E, Levitskii V S, Shapovalov V I *et al* 2016 *Glass Phys. Chem.* **42** (*in print*)
[3] Shapovalov V I 2015 *Nanoingeneria* **11** 22–34
[4] Särhammar E, Nyberg T and Berg S 2015 *Sur. Coat. Technol.* **279** 39–43
[5] Barybin A A and Shapovalov V I 2007 *J. Appl. Phys.* **101** 054905
[6] Barybin A A, Zavyalov A V and Shapovalov V I 2012 *Glass Phys. Chem.* **38** 396–401

Magnetron sputtering as a method of thin-film catalyst development for electrochemical sensors

E A Medvedeva

FSUE "SPA"Analitpribor", Smolensk, Babushkin st., 3, Russia

E-mail: elenamedvedeva78@mail.ru

Abstract. The aim of this work was to develop a thin-film Pt/C catalyst on the fluoroplastic substrates by means of the magnetron sputtering method in order to use as reference and working electrodes of electrochemical cells.

Nowadays a method of magnetron sputtering of materials has obtained a wide spread occurrence owing to a growing demand for different high quality coatings such as: wear-resistant coatings, coatings with low friction factor, corrosion-resistant coatings, coatings with given optical and electrical properties, etc. [1].

The aim of the present work was to develop a thin-film Pt/C catalyst on the fluoroplastic substrates by means of the magnetron sputtering method in order to use as reference and working electrodes of electrochemical cells. The choice of nanocomposite Pt/C thin-film electrodes is related to the fact that the materials on the basis of amorphous carbon have unique physical and chemical properties [2], and an insertion of metals nanoclusters into the matrix of amorphous carbon allows to control the electrochemical activity of Pt/C.

The advantage of magnetron sputtering according to the properties of nanocomposite catalytic layers is a possibility to obtain thin-film coatings from two and more materials in any ratios, adjust the thickness of the sputtered layer and therefore its conductivity. In addition, during magnetron sputtering a high purity of catalyst is provided due to lack of foreign impurities in the sputtering chamber.

Works concerning magnetron application of thin-film Pt/C catalysts have been performed before [3]. The task of the present work was to obtain Pt/C catalytic layers with the required characteristics:

– high porosity for gas diffusion;

– high conductivity ($R \sim 60\ \Omega$);

– possibility of synthesis of the composite on the fluoroplastic substrate with minimum thermal load;

– maximum developed surface of the obtained composite, as characteristics of its catalytic activity;

– maximum purity of the composite due to high sputtering rate of the target, the higher the rate is, the purer the film is.

The substrate and the conditions in the deposition chamber have an effect on the properties of the obtained composite in the first place. By means of the appropriate parameters of the substrate (roughness, porosity, structure), vacuum control, formation of definite gaseous medium in the chamber and control of plasma combustion conditions, sputtering control goes with a good reproducibility.

The thin films of Pt/C composite obtained on the equipment "Elato 600-P" ("Izovac") equipped with DC magnetron and substrate holder rotation system (figure 1).

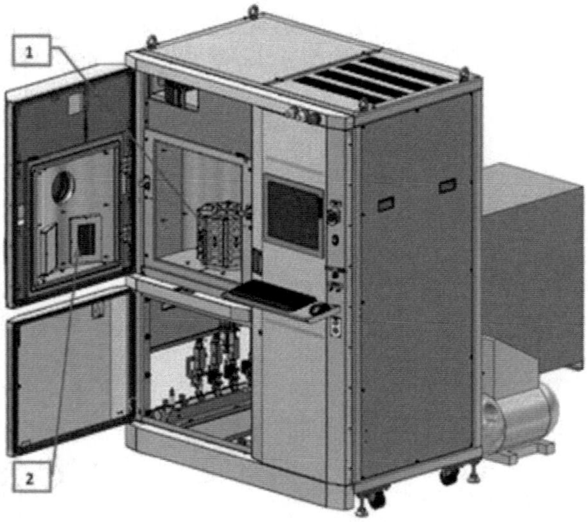

Figure 1. "Elato 600-P" external view: 1 – substrates rotation system; 2 – magnetron sputtering system.

Films deposition was carried out in argon atmosphere at a pressure ~10^{-1} Pa. Sputtering mode: U = 500 V, I = 200 mA, t = 20 min. Layers growth rate was 30–35 nm/min.

The selected parameters of the sputtering mode of nanocomposite electrodes have provided the specified resistance of the electrodes (35–40 Ω at sputtering time 20 min at a distance between the probes 1 cm).

SEM research of catalyst surface has shown a sufficient porosity of the synthesized composite (figure 2), as evidenced by gas permeability test of the obtained membrane. In addition, the layer thickness ~ 500 nm provides hydrophobicity of the obtained layer at good hydrophilic properties of the membrane surface.

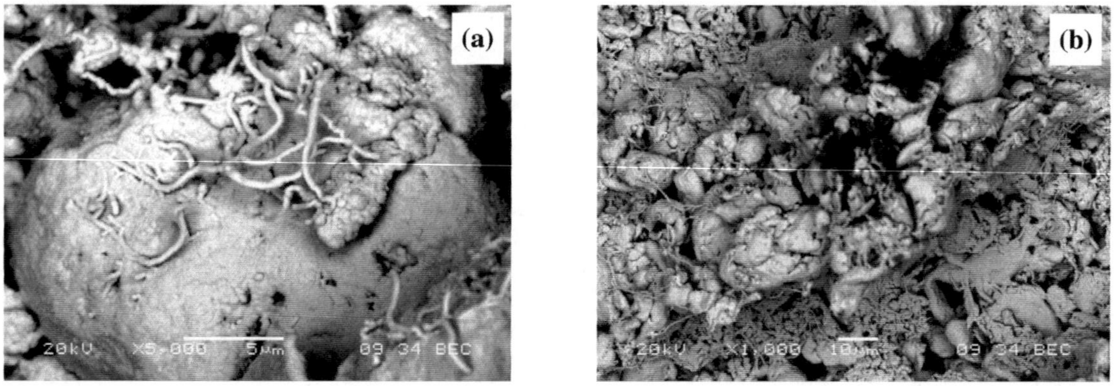

Figure 2. Microphotographs of electrodes surface: (a) – resolution 5 μm; (b) – resolution 10 μm.

By means of additional research of the composite surface with energy dispersive spectrometer Princeton Gamma-Tech, purity of the obtained thin film has been proved (table 1, figure 3).

Table 1. Percentage ratio of platinum and graphite in two arbitrary points.

Element	Wt %	At %
Pt	92.75	44.07
C	7.25	55.93
Total	100.00	100.00

Element	Wt %	At %
Pt	92.48	43.10
C	7.52	56.90
Total	100.00	100.00

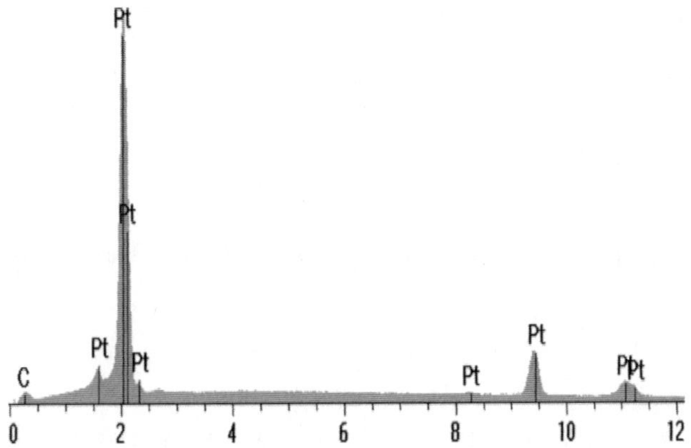

Figure 3. Research results of Pt/C composite obtained by means of energy dispersive spectrometer Princeton Gamma-Tech.

When installing the obtained electrodes into the electrochemical sensor (ECS) designed for determination of hydrogen sulfide level in the air of working area the sensor showed its high selectivity, including lack of sensitivity to carbon monoxide (table 2).

Table 2. Sensors cross sensitivity.

Test gas	Concentration (ppm)	Sensors readings	
		For CO (ppm)	For H_2S (ppm)
Chlorine	100	0	0
Carbon monoxide	100	100	0
Hydrogen sulfide	30	0.5	30
Nitrogen dioxide	30	0.3	0

The investigated long-term stability of ECS has shown, that during six months of operation sensitivity reduction was less than 1 %. ECS sensitivity 0.45 $\mu A/mg/m^3$ is comparable with sensitivity of commercial samples. Cross sensitivity to hydrogen chloride, sulfur dioxide, formaldehyde, ethylene oxide, ammonia and other gases is equal to zero.

References

[1] Kelly P J and Arnell R D 2000 *Vacuum* **56** 159–72

[2] Sleptsov V V, Kuzin A A, Ivanovsky G F, Elinson V M, Gerasimovich S S, Kondrashov P E and Baranov A M 1991 *J. Non-Crystalline Solids* **136** 53–9

[3] Medvedeva E, Baranov A and Somov A 2016 *Sensors and Actuators B: Chemical* (*in print*, DOI: 10.1016/j.snb.2016.02.104)

Journal of Physics: Conference Series **729** (2016) 012008 doi:10.1088/1742-6596/729/1/012008

Formation of deep mesa-structures on SiC using fluoride plasma

A V Afanasiev, V A Ilyin, A A Romanov, A V Serkov and D A Chigirev

St. Petersburg Electrotechnical University "LETI", Professora Popova str. 5, St. Petersburg, 197376, Russia

E-mail: cmid_leti@mail.ru

Abstract. A process of deep profiles formation on silicon carbide using ion-plasma etching in fluoride plasma is discussed. An industrial equipment «Caroline PE 15» (Russia) with ICP plasma source was used in experiments. The obtained experimental samples are perspective in power electronics applications.

1. Introduction

Silicon carbide shows an extreme hardness, which makes its mechanical and chemical processing. Ion-plasma processing in vacuum provides a new level for problem solving in many technological tasks, which is determined by selectivity, homogeneity and anisotropy of etching process (at high etching rate), as well as a possibility for generation of materials with various electrophysical properties.

At present time one of the principal methods for semiconductor structures microprofiling is reactive ion-plasma etching (RIE). In microstructure formation this technology provides sufficiently high etching rate, possibility for realization of required profile of elements and low density of defects at subsurface area of the structures.

This article is devoted to microprofiling of silicon carbide in fluoride containing plasma, revealing and optimization of processing regime in formation of deep mesa-structures using industrial RIPE-equipment «Caroline PE 15» (Russia) (figure 1).

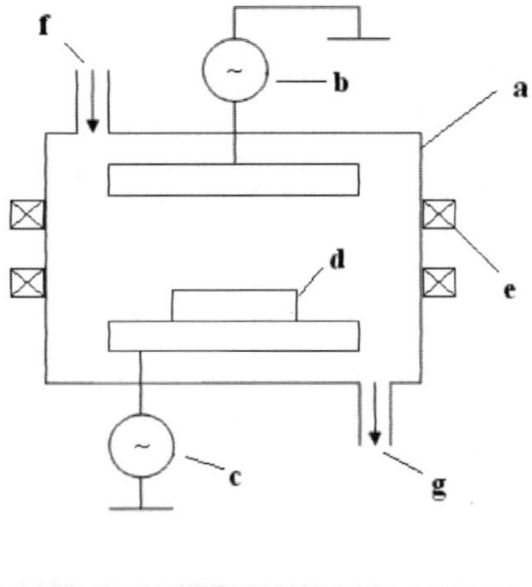

Figure 1. Scheme of RIE equipment «Caroline PE 15»: a – reactor chamber; b – ICP; c – cooled HF-electrode; d – processed sample; e – spools of magnetic system; g – flange for gas delivery; f – flange for pumping.

Journal of Physics: Conference Series **729** (2016) 012008 doi:10.1088/1742-6596/729/1/012008

2. Experimental

Complex silicon carbide surface cleaning procedure was completed at a preparatory stage of work using liquid treatment with $H_2SO_4 + H_2O_2$ mixture at 60…70 °C. Then, multilayer metal deposition of Ti–Cu–Ni was performed using vacuum equipment for magnetron deposition «KONT» (Russia). Titanium was used in order to increase an adhesion, copper – as a general mask and nickel – as a protection metal for copper, to prevent oxidation of copper contacts.

In the next step photolithography was made using metallization as a mask, and after that deep mesa-structures were formed by RIE method. In experiments a mixture of sulfuric hexafluoride with oxygen was used. RIE process is an ion-stimulated one, because radicals and ions of SF_n are capable for reacting with substrate surface. The processes in plasma and on the silicon carbide substrate surface could be schematically represented as follows:

$$SF_6 + e^- \Rightarrow SF_{6-n}^{\ n+} + nF + e^-;\ n = 0...6;$$

$$Si + 4F \Rightarrow SiF_4\uparrow;$$

$$Si + SF_4^{2+} \Rightarrow SiF_4\uparrow + S\downarrow;$$

$$C + 4F \Rightarrow CF_4\uparrow.$$

In the process of etching serum is precipitated on the substrate surface. It could be removed by physical nebulisation or by adding oxygen into the working gas mixture:

$$S + O_2 \Rightarrow SO_2.$$

Adding oxygen increases the rate of SiC etching, since oxygen reacts with carbonium yielding volatile compounds which prevents formation of polymer films on the surface:

$$C + O_2 \Rightarrow CO_2\uparrow.$$

Generation of chemically active particles in plasma causes simultaneous development of various processes, such as etching, polymerization, gas-phase reactions. Furthermore etching rate depends practically of all parameters of the process: etching gas pressure, HF-generator power, temperature and working gas flow velocity.

In order to reveal optimal etching rate values, silicon carbide surface quality and wall slope angle a series of experiments were carried out. A fixation of delivery rate of $SF_6 + O_2$ mixture and spool current (I_s = 1,2 A), while power applied at upper (150…700 W) and lower (100…300 W) HF-electrodes varied (figure 2).

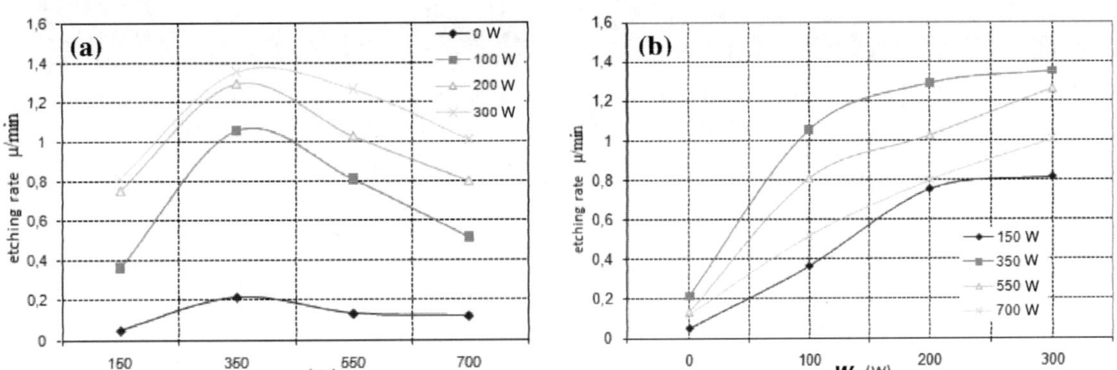

Figure 2. Dependence of etching rate for silicon carbide from: (a) – power at the upper HF-electrode (W_{up}) at various power of lower HF-electrode values; (b) – power at the lower HF-electrode (W_{lo}) at various power of upper HF-electrode values.

Journal of Physics: Conference Series **729** (2016) 012008 doi:10.1088/1742-6596/729/1/012008

Figure 2(a) shows that under fixed power value at lower HF-electrode, increase of etching rate is observed along with increase of discharge initiation power, which could be explained by the increase of dissociation grade and ionization of working gases, and consequently – concentration of active particles.

There is a range of optimal power values, where etching rates are maximal. When power is increased, a free-path length of particles decreases, which makes delivery of active particles released in the dissociation process and responsible for etching, hard to reach the substrate. Under power lower than the optimal one concentration of above particles decreases.

Figure 2(b) shows, when power of lower HF-electrode is increased the intensity of the ion sputtering surface increases. Etching rates of mask layers in RIE were: for nickel – 15…17 nm/min, copper – 30…40 nm/min, titanium – 150 nm/min. These values provide etching rate ratio values within ~ 40…50.

3. Results and discussion

An optimal regime for deep silicon carbide etching was experimentally developed and tested:

– power at upper HF-electrode W_{up} = 350 W;

– power at lower HF-electrode W_{lo} = 200 W;

– spool current I_s = 1,2 A;

– gas mixture of $SF_6 + O_2$ pressure P = 1,5 Pa.

Fixed auto-offset voltage was –120 V. In figure 3 SEM images («Quanta Inspect», USA) of RIE processed silicon carbide are presented.

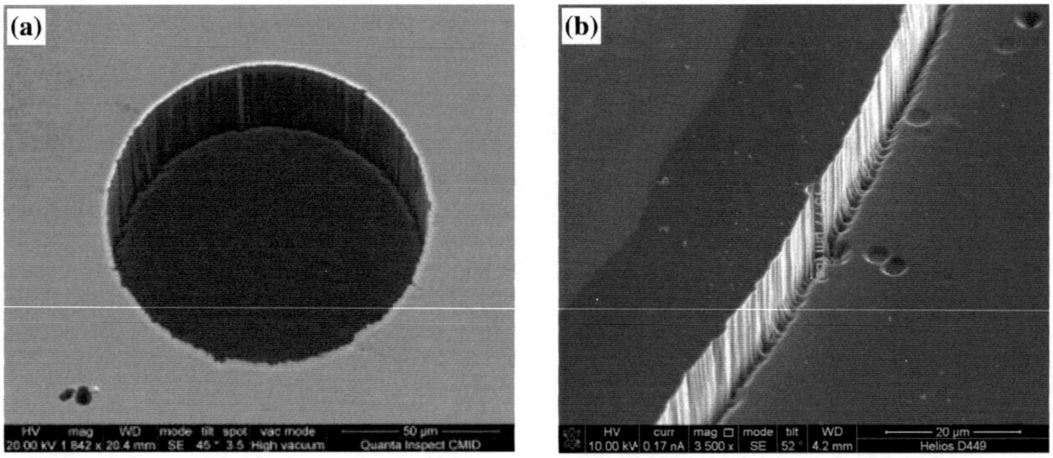

Figure 3. SEM-images: (a) – topological image formed at the surface of silicon carbide crystal after RIE; (b) – side wall of mesa-structure formed by RIE on silicon carbide.

Etching depth of ~15…16 µm is obtained for good quality of surface and side walls. Side-wall angle is ~ 90°.

4. Conclusions

Silicon carbide mesa-structures were formed using RIE method and RIE equipment «Caroline PE15» (Russia). The mesa-structures are perspective for use in devices of power electronics.

Optimized technological parameters of RIE process enabled a deep profiles to be formed in silicon carbide for preparation of semiconductor instruments. It has been shown that metals used as masking materials show sufficiently high selectivity, providing formation of deep profiles.

The mechanism of formation of the droplets on the electrodes under the impact of the high power density streams

V D Goncharov[1], K S Sorokin[1], R V Yashkardin[1] and E M Fiskin[2]

[1] Saint Petersburg Electrotechnical University "LETI", Laboratory of "Electric impulsive technologies", Russia
[2] Irkutsk state technical university, Russia

E-mail: vdgoncharov@rambler.ru

Abstract. The article contains a brief description of functioning principles of the device for creating ultradispersive powders of metals under the impact of electrically charged streams with power density about 10^8 W/cm^2. The results of atomic forced microscopy (AFM) measurements of surfaces of the electrodes exposed to dispergate the microdroplets, which allowed to study the droplets formation steps are presented. The results of AFM surfaces of the substrate surfaces to be inflicted by the dispergated droplets are presented. The dependency of the particles sizes on the distance between the electrode and substrate allows to consider the main mechanism of division the dispergated from the electrode surface droplets the Rayleigh instability.

Nanomaterials and coatings based on these materials tend to find increasingly expanding cases of use in different areas of science and technics [1, 2]. It is widely known, that nanoparticles may act as the catalysts, the coatings of these materials may be used for disinfection, or to dissipate the electromagnetic waves or as a structural strengthening phase. Every year the areas of applying of those materials increase. Though the most of the methods of their production are based on the chemical reactions, while using ones, the complicated fractal structures of the materials appear. Further those structures might be delivered to the surfaces to be used with the nanoparticles. But they could still tend to aggregate there and in consequence of that, loose their useful properties. The alternative methods of producing ultradispersive (smaller than 100 nm) particles, are the physical ones, based on the influences of powerful streams of energy on dispergated materials, are still in the phase of development [3–5].

One of the examples of the physical methods is the device for producing the ultradispersive materials is being realized in the laboratory of "Electric impulsive technologies" SPb ETU [6, 7].

A brief description of the device's working principles:

1. The addictive sparking discharge of the atmospheric pressure, which is initialized by impulsive high voltage transformer through high voltage addictive electrodes, initializes the ignition of the main discharge. This discharge initializes the release of the energy, previously stored in the capacitive batteries, which are connected to the main electrodes with the co-axial cable;

2. After the ignition, the discharge exists in the evaporations of the electrodes material. Under the influence of power, appearing while the magnetic field of currents on the electrodes and plasma, the discharge starts to move along the rails from the point of connection of the electrodes;

3. During a short period of time (100 μs) the huge energy input (1500 J) with high power densities (10^8 W/m^2) takes place in the discharge plasma and, consequently, high temperature of plasma and electrodes and big currents (up to 50 kA) appear. The average time of the contact of discharge and the point on the electrode is no longer than 10 μs. During this time the surface of the electrode heats up, melts, partly evaporates and dispergates (part of the material leaves the electrode surface as a microdroplet phase);

4. The droplets, moving influenced by vapor pressure from the evaporated electrodes and the electric field, rapidly start to move to the substrate. The droplets, that reached the substrate and held on there, get frozen as discrete particles.

The droplets formation lasts for couple microseconds. Its size reaches couple microns, the temperature – about 5 000 degrees. The process of formation of these droplets is very complicated. The ways of experimental studying of the process in such conditions are also very complicated and expensive [8]. Though they can't give the answers about the microdroplet formation process, which is key in our case.

Having in mind the complexity of theoretical and experimental research of the processes, we used the method of indirect measurements. In terms of this method we judged about the processes to take place on the electrode by the trails, left by running discharge. The problem is that the discharge moves along the extended electrodes surface making it melted on couple microns deep. When the discharge leaves the area on the electrode, it rapidly cools down and many droplets, which were in the process of formation, get frozen as they were in the area of the discharge. That's why to find out the physical mechanisms of the droplets formation, it's necessary to measure the electrodes surfaces, finding out the characteristic areas on it and also to research the changes in the particles dispersions on the substrate depending on the distance to the electrodes.

Sizes of the particles, which form on the electrodes and get caught on the surface of substrate change from couple nm to couple um. That's why to measure the surfaces of electrodes and substrates, we used atomic-forced microscope (AFM) Certus in tapping mode.

The measurements of surfaces of electrodes and substrates with the particles coatings, received while using different modes of working of the technological device (voltage, capacitance, gap between electrodes and substrates, etc.). Firstly, the surface of the unused electrode was measured (figure 1).

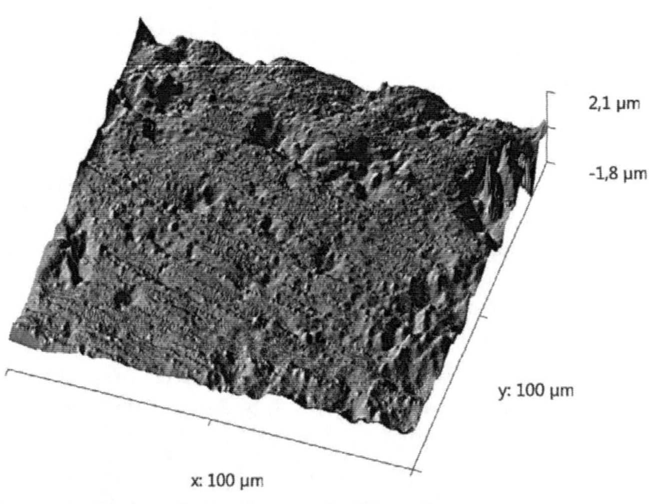

Figure 1. The surface of unused electrode measurement result.

It's easy to see that the surface is mostly flat. There're no significant altitudes over or under the main surface, even of the sector of 100 μm. The height of arbitrary located irregularities does not exceed 3–4 μm.

Scanning of the electrodes in different areas after using them appeared that for different stages of the discharge movement the surfaces of the electrodes have different structure (figures 2, 3) [9].

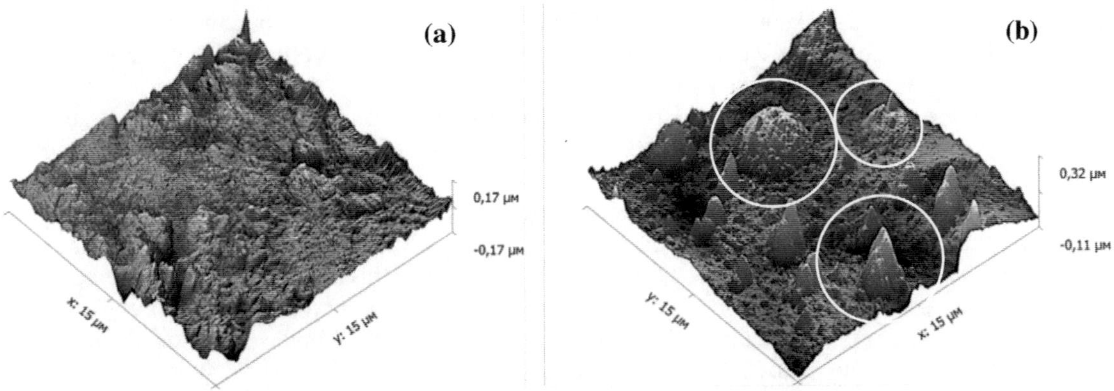

Figure 2. The images of the surfaces of the electrodes after their use in different points: (a) – in the middle of the discharge trail (the surface is still pretty flat, though slightly more rough than unused); (b) – at the end of the trail (we can see the droplets forming on the surface of the electrode).

Figure 3. The photo of the electrode with the discharge trail.

For the images, created with scales of 100×100, 50×50 and 15×15 µm, different characteristic regions were allocated, where surface roughnesses had same diameter and different altitude. Those areas are marked and zoomed in figure 4.

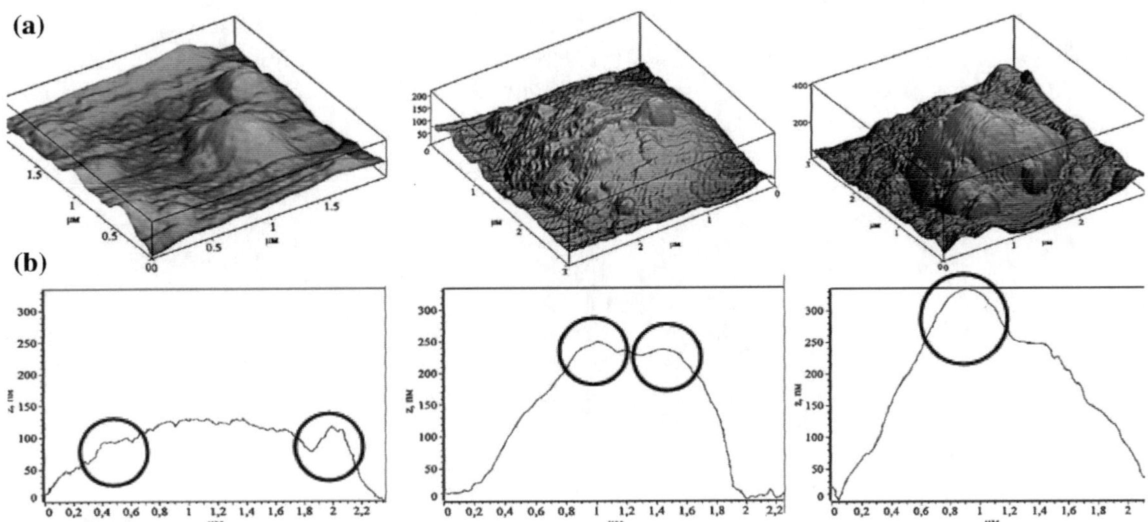

Figure 4. The images of forming of the droplets on the surface of the electrodes on different stages of forming: (a) – three-dimensional pictures; (b) – two-dimensional pictures of the droplets, where we can see the forming of the secondary, much smaller droplets.

The scales for all coordinates kept same for all three cases. Obviously, these roughnesses have a shape of a piece of sphere, about the same diameter, and different height. We consider these spheres as frozen on different stages of forming microdroplets.

The droplets of a much smaller size also can be found on the measured surface. From the surfaces of big droplets, sized about 2 μm the droplets of 200 nm separate. Most likely this process is a result of the Rayleigh instability.

The results of measuring the surfaces of substrates, covered by particles for the different cases of heights over the electrode are shown on the figure 5. The material of the electrodes was steel, the voltage was 3 kV, capacitance 500 μF. Obviously, the created particles depending on the substrate removal from the electrodes tend to have different sizes and distance between each other. This gives the opportunity to say, that while reaching on to the substrate the droplets keep on dispergating, becoming smaller while the removal from the electrodes.

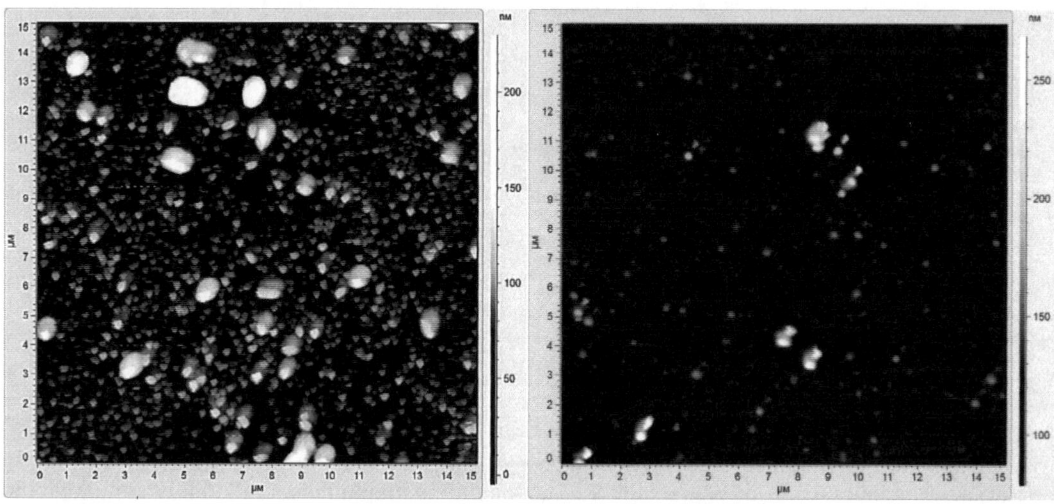

Figure 5. The results of measurements of the surfaces of the substrates at different distance from the electrode (left 5,5 cm, right – 7,5 cm, at 3 kV and 500 μF).

Particularly, on the distance of 5,5 cm from the substrate, most of the droplets have a size about 150–180 nm, the size of particles is going to be 100 nm. At the 7,5 cm from the electrode the size decreased to 50–120 nm (figure 6). The number of particles grows significantly. Though the summary volume keeps almost unchanged.

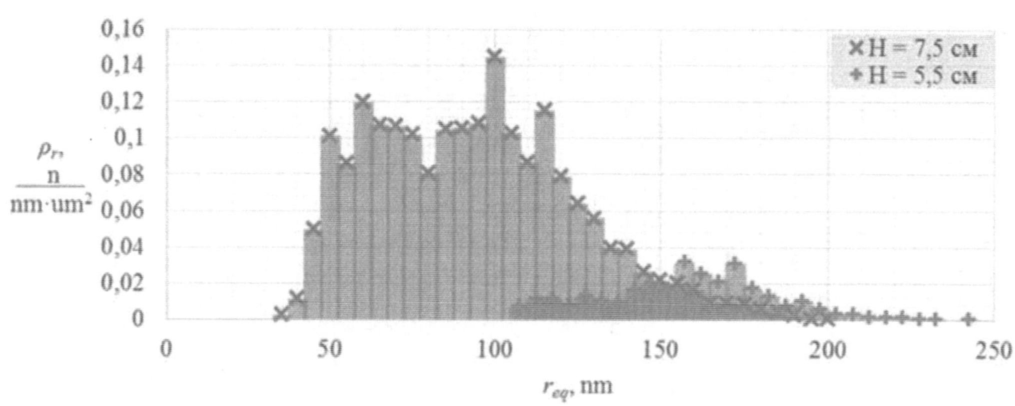

Figure 6. The results of mathematical processing of the pictures of the substrate on different distances from the electrodes.

Journal of Physics: Conference Series **729** (2016) 012009 doi:10.1088/1742-6596/729/1/012009

Conclusions

The measurements created with AFM of the electrodes of the system for producing ultradispersive coatings, allowed to investigate and capture the phases of forming of the droplets under the affection of powerful charged energy streams.

The results of AFM measurements created using different parameters of the technological device and on the different distance from the electrodes allow to claim, that the division of the droplets still continues while their flying in the plasma of the vapors of evaporated material of the electrodes.

On the assumption of the produced researches, we can suggest that the main mechanism of the droplets formation on the electrodes and their division in the discharge plasma is Rayleigh instability.

References

[1] Kostrin D K and Lisenkov A A 2016 *Materials Science Forum* **843** 278–83

[2] Vetrov N Z, Kostrin D K, Lisenkov A A and Popova M S 2015 *Journal of Physics: Conference Series* **652** 012032

[3] Zabrodsky A G, Gurevich S A, Kozhevin V M, Astrova E V, Nechitailov A A, Sresely O M, Terukov E I and Compon M E 2007 *International Scientific Journal for Alternative Energy and Ecology* **2**

[4] Russkih A G, Oreshkin V I, Labetskiy A Yu, Tchaikovsky S A and Shishlov A V 2007 *Journal of technical physics* **5** 35–40

[5] Goncharov V D and Samsonov D S 2015 *Journal of technical physics* **5** 37–42

[6] Samsonov D S, Goncharov V D and Fiskin E M 2011 Patent RU 2471884

[7] Goncharov V, Samsonov D, Fiskin E and Kalinin S 2014 *Proceedings of the international scientific colloquium Modelling for electromagnetic processing (Hannover)* (Moscow: Spektr Publ. House) pp 277–81

[8] Zhykov B G, Kurakin R O, Sakharov V A, Bobashev S V, Ponyaev S A, Reznikov B I and Rosov S I 2013 *Letters to JTP* **23** 58–62

[9] Goncharov V D, Kalinin S A and Yashkardin R V 2015 *Proceedings of Saint-Petersburg state electrotechnical university «LETI»* **5** 3–7

Journal of Physics: Conference Series **729** (2016) 012010 doi:10.1088/1742-6596/729/1/012010

Producing multilayer composites based on metal-carbon by vacuum ion-plasma method

S R Shekhtman and N A Sukhova

Ufa State Aviation Technical University, Russia

E-mail: shex@inbox.ru

Abstract. The possibility of changing the coating properties and perspective of forming a compound having the unique mechanical properties, high hardness and high corrosion resistance is of particular interest to multilayer coatings based on metal-carbon composition. To investigate the mechanisms of formation of coatings and to demonstrate technological possibilities synthesis technology of multilayer composites, the titanium cathode and the silicon-graphite cathode were made. Coatings microhardness analysis showed that in forming the multilayer structure, microhardness increases by 15–20 %, and when forming a multilayer composite, microhardness increases by 60–65 %.

In most cases, when the parts are exposed to an aggressive environment, such as wear abrasion, corrosion and erosion, exposure to high temperatures, the destruction begins from the surface. To protect the surface, a method of vacuum-plasma deposition of coatings is most prevalent. This method is characterized by the possibility of obtaining surface layers with special physical and mechanical properties, and helps to protect the parts during the entire operating life. The suggested method has an almost complete universality of the technological process, high productivity, high reproducibility of parameters on each operation and almost complete lack of environmental problems [1–3].

There are several main reasons hindering the widespread implementation of ion-plasma coatings in friction units of machines and mechanisms, in particular aviation ones.

First, for friction units of aircraft components, we need to provide wear resistance and corrosion protection, which is not always achieved when the protective layer thickness is several micrometers (3...5 μm) produced by traditional technologies.

Secondly, it is necessary to ensure sufficient fatigue strength of the parts with coating. It can be achieved by a reduction of residual stresses. Therefore, we need to increase the thickness of the coating, to create therein transition layers or multilayer structure [1].

With this in mind, we carried out studies on the creation of multilayer coatings of Ti–C system. In this case, multilayering provides a smooth reduction of residual stresses on thickness, a reduction of porosity and defects in the coating, an increase the integrated thickness of coating without sacrificing operational characteristics and an ensuring a smooth transition from soft base to solid functional layer [2, 3].

The possibility of changing the coating properties and perspective of forming a compound having the unique mechanical properties, high hardness and high corrosion resistance is of particular interest to multilayer coatings based on metal-carbon composition.

For the deposition of coatings with improved properties, the technology of coating multilayer composite has been developed in relation to the modernized industrial plant NNV 6.6-I1, which includes: a preliminary surface preparation; ion cleaning of the surface; applying a multilayer composite (carbon–metal).

Plant lay-out for coating a multilayer composite is shown in figure 1. The preliminary surface preparation is used to remove from the surface different kinds of pollution. Then the parts are placed in the vacuum chamber and pressure is generated. Preliminary ionic purification is carried out by a non-self-maintained high-current diffusion discharge generated by plasma source PSIC, the use of which gives the additional acceleration of ions and acceleration of diffusion processes in the near-surface layers of the treated parts.

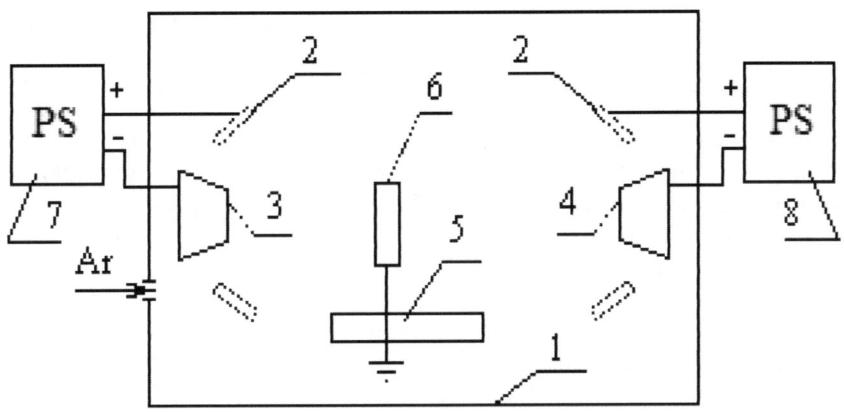

Figure 1. Plant lay-out of NNV 6.6-I1 modernized for applying the multilayer composites: 1 – vacuum chamber; 2 – anode; 3, 4 – cooled cathodes made of sprayed material, titanium and graphite-doped silicon, respectively; 5 – table; 6 – part; 7, 8 – power supply.

The next step of technology is a cleaning the surface by the metal ions. It is carried out in a vacuum chamber at a certain pressure in an inert gas argon (Ar) atmosphere. The accelerating voltage is not less than 1100 V is applied to the substrate at an arc current of 40–50 A. Duration of procedure is 1–2 min till the termination of microarcs occurrence. Cleaning also allows preheating the parts before coating.

Step of technology related with the formation of the multilayer composite (carbon–metal) occurs during the serial deposition of layers of metal and carbon from plasma generated by electric arc plasma sources arranged at an angle to the substrate surface.

To investigate the mechanisms of formation of coatings and to demonstrate technological possibilities synthesis technology of multilayer composites, the titanium cathode (the Russian brand VT-1.0) and the silicon-graphite cathode were made. In the process of coating deposition, substrate evenly rotates on its axis and on the axis of the table together with the plant.

To determine patterns of synthesis, the samples were coated by the following types: a single-layer Ti–C coating, a multilayer Ti–C–Ti coating and the multilayer composite. A layer of titanium was the first and the last layers of each coating.

Investigations on adhesion strength of multilayer composites and multilayer coatings showed that the coatings have excellent adhesion to a base material. It was obtained by the study of the diamond pyramid's print scaled up by using an optical microscope. Lack of cracking of the material with coating near the impact zone of the indenter is proof of adhesive strength.

Coatings microhardness analysis showed that in forming the multilayer structure, microhardness increases by 15–20 %, and when forming a multilayer composite, microhardness increases by 60–65 %.

Corrosion resistance of samples with coatings was determined by gravimetric method. Corrosion resistance of the multilayer composite (Ti–C) is higher by 20–22 % than that of samples with multilayer coating and by 25–30 % than that of the uncoated samples (see table 1).

Table 1. The corrosion rate of samples with coatings.

Type of coating	Mass, g		Square of samples, m^2	Corrosion rate, $g/m^2 \cdot h$
	before experiment	after experiment		
Initial	4,39355	4,38625		0,0222379
Multilayer coating	3,76240	3,75635	0,0015198	0,0184301
Multilayer composite	4,39755	4,39245		0,0155361

Investigations of the electrode potential to determine the effectiveness of the multilayer composite Ti–C as protective coatings were carried out by the conventional method of observing the changes of the samples potentials in time. We used the block of universal voltmeter with a range of 7–35 V. The electrolyte was a 5 % solution of NaCl. Reference electrode was silver chloride (AgCl) half-cell, immersed in a solution of KCl (saturated).

The electrode potential of multilayer coating applied to samples of steel is higher by 20–25 % than the electrode potential of the sample coated with submicrocrystalline (SMC) structure (steel of the Russian brand 13H11N2VMF). Consequently, the surface of the multilayer composite Ti–C is more passive, which indicates its higher corrosion resistance. This can be explained by an increase of the number of boundaries between the layers. It inhibits corrosion processes at the boundary between medium and metal and prevents the penetration of external corrosive environment deep into coatings (figure 2).

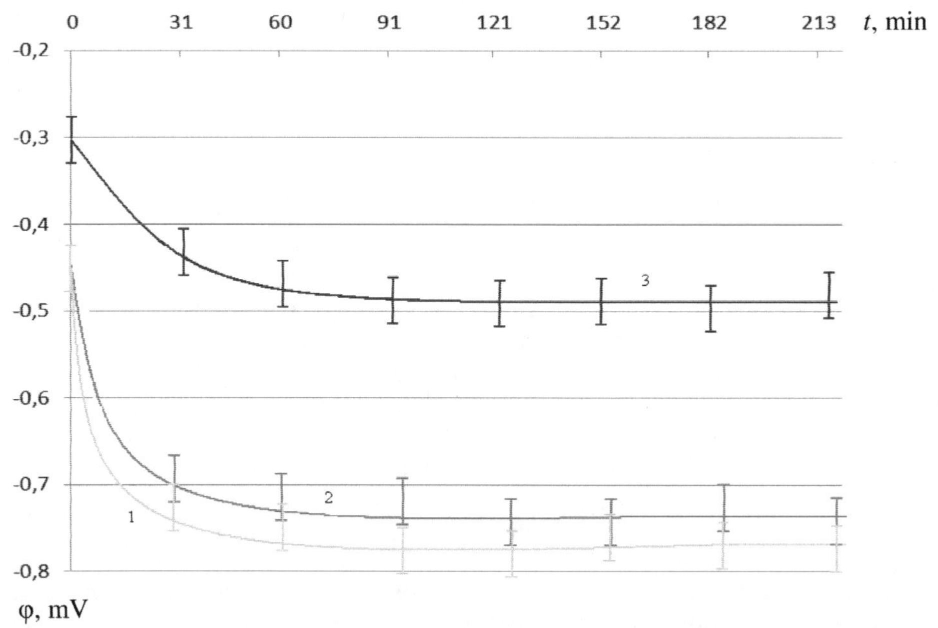

Figure 2. Electrode potential. Substrate 13H11N2VMF: 1 – initial state; 2 – multilayer coating; 3 – multilayer composite.

Analysis of the samples with a multilayer composite Ti–C, obtained by vacuum ion-plasma method showed that the coatings deposited using plasma assisting provide higher protection against corrosion.

Furthermore, improved corrosion properties can be explained by increased density of the coating, restructured coatings and reduced residual stresses in the coating.

It was found that in comparison with a multilayer coating, the microhardness of coatings with SMC structure is higher by 30–50 %, electrode potential is lower by 20–25 %, the corrosion resistance of the multilayer composite (Ti–C) is higher by 20–22 % than that of samples with multilayer coating and by 25–30 % than that of the uncoated samples that is associated with an increase in the content of carbide and titanium carbosilicide in a multilayer composite.

References

[1] Budilov V V, Kireyev R M and Shekhtman S R A 2007 *The technology of vacuum ion-plasma processing* (Moscow: MAI)
[2] Mukhin V S and Shekhtman S R 2007 *Vestnik UGATU* **1** 84–91
[3] Budilov V V, Mukhin V S and Shekhtman S R 2008 *Nanotechnology of surface treatment of parts by vacuum ion-plasma methods: physical fundamentals and technical solutions* (Moscow: Nauka)

Journal of Physics: Conference Series **729** (2016) 012011 doi:10.1088/1742-6596/729/1/012011

Change of energy state in the surface layer at the stages of vacuum ion-plasma treatment process

V V Plikhunov and L M Petrov

JSC "National Institute of Aviation Technologies", Moscow, Russia

E-mail: info@niat.ru

Abstract. Change of energy state in the surface layer under treatment at the stages of vacuum ion-plasma process is discussed. It is demonstrated that the most informative way to obtain data on the surface condition after the action of plasma streams is a method of contact potential difference used for evaluation of the energy state of the surface. Different examples of the application of this method to evaluate the surface energy state and to determine effectiveness of plasma streams action are presented.

Vacuum ion-plasma (VIP) treatment is one of the widely used technological methods for surface strengthening. This method makes it possible to form external, internal and combination coatings employing a wide range of plasma stream effects.

Provision of the necessary level of performance properties during VIP treatment is achieved by changing the structure of the work surface of structural metal materials under the effect of high-energy gas-metal plasma streams. The influence of plasma streams involves a change in the energy state of the surface layer, which is an estimation characteristic of effectiveness of plasma streams effects as well as a quality criterion of the generated coatings. The importance and the necessity for an investigation into the energy state of the surface layer during surface treatment have been mentioned in [1, 2], however, at present the change of energy state, as a rule, is never taken into consideration. The reason is first of all associated with the lack of reliable and high-quality instrumental equipment that enables us to evaluate the change of energy state of the surface at different treatment stages.

Analysis of the methods used for control of the energy state of the surface layers has permitted to determine that the most informative and on-line method of evaluating the change in the energy state of the surface layer is a method of contact potential difference. According to the value of the surface potential of pre- and post-plasma stream effects we can estimate the effectiveness of these effects on the work surface. Devices that implement the given method are characterized by high sensitivity to external electromagnetic action, which complicates assurance of stability, when measuring surface potential values [3].

The analyzer of the surface energy-state (ASES) [4] designed and produced in the National Institute of Aviation Technologies (NIAT) permits realization of the non-destructive testing of the energy state of the surface of structural metal materials. This method includes the use of technique for measurement, a measuring device, and a procedure for operation with the measuring device and also a procedure for the estimation of measurement results. The principle of this device operation (figure 1) is based on the fact that difference of contact potentials of the specimen and the measuring electrode, which are in contact at the moment of their separation, forms a capacitor with an electric charge Q in the specimen – measuring electrode system. The feedback electrical circuit provides for conservation of the charge value Q = const through generation of the compensating voltage Y without any additional voltage sources. With constant speed of the measuring electrode removal, the value of compensating voltage will also increase in direct proportion to time t from the moment of separation of the specimen and the measuring electrode. Thus ASES device allows us to obtain a graph of linear function $V(t) = At$, where A is a constant that determines an angle of inclination α of the straight line $V(t)$ in Cartesian coordinates V–t. It should be noted that constant A is directly proportioned to the value of charge Q.

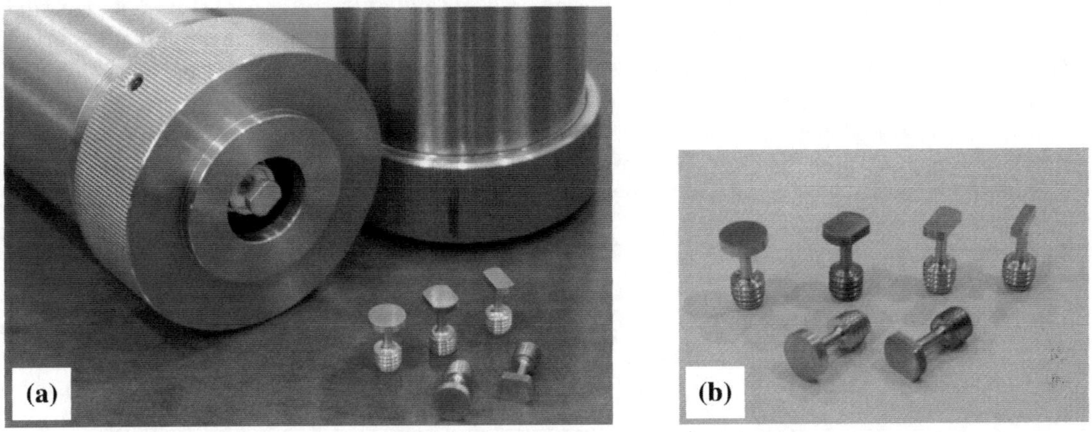

Figure 1. The analyzer of the surface energy-state (a) and replaceable sensors (b).

The energy state of the surface of specimens is compared either from the value of angle α (in this case $Q \sim \text{tg}\ \alpha$) or from the value of voltage V defined at a given time value t (for instance, t = 200 μs). In the latter case we will call voltage V as a reduced surface potential V_p underlying thereby dependence of its values on the selected time period and technical features of the ASES design including size, material and measuring electrode travel speed.

Taking measurements by means of the ASES device in the same conditions, it is possible to compare the energy state of the surface of specimens made in different materials under different types of treatment. Figure 2 presents examples of definition of the reduced surface potential V_p of a 40H steel specimen during VIP treatment with the aid of ASES device.

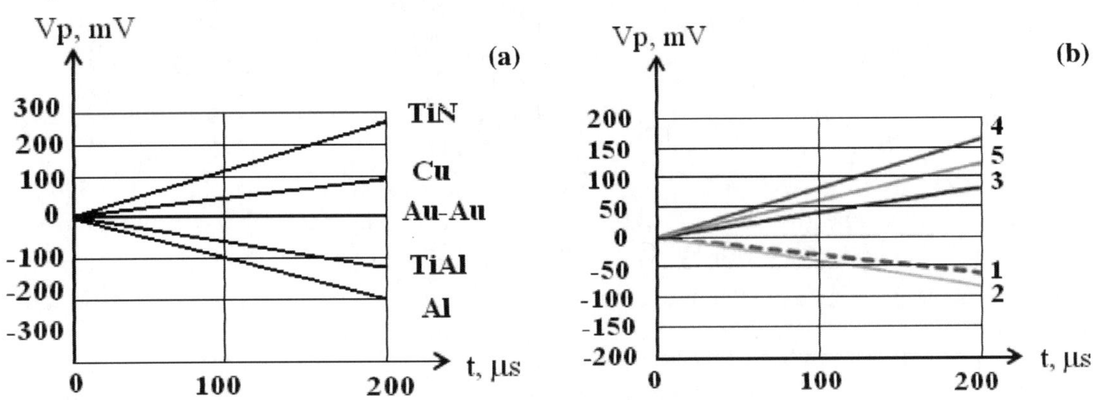

Figure 2. Behaviour of V_p values of coatings relative to the sensor gold-plated electrode (a) and dependent on the modes of electronic heating with assisted Ti deposition (b): 1 – initial condition; 2 – 300 °C (Ti –1,2 mass %); 3 – 400 °C (Ti – 2,4 mass %); 4 – 500 °C (Ti – 4,8 mass %); 5 – 600 °C (Ti – 5,4 mass %).

The study of variations in V_p values depending on the surface condition of the structural metal materials enabled us to establish the regularities of their change subject to tooling influence as well as different types of surface contamination in the shop-floor atmosphere (dust, oil vapors, coolant, etc.). It has been found that the variations of the reference value V_p indicate that under process effects considerable surface barrier layers are formed, which inhibit generation of high-quality coatings with high level of operational properties [5].

The pursued investigations have permitted to determine effectiveness of the process effects of plasma streams at VIP stages and to substantiate modes of their implementation. It has been found out that in the course of ion etching of 40H steel by argon gas plasma in the mode of semi-self-maintained gas discharge $V_{ref} = -500$ V and the value of V_p changes according to the time of ion stream action. In this case the beginning of ion etching process within one minute is characterized by transition of V_p value from the negative domain to the positive domain (figure 3(a)), which indicates that contaminating adsorbed layers and oxides have been removed from the work surface. The basis for selection of the etching mode is the presence of maximum activation values at the surface. However, the maximum value of V_p is observed under 20 min ion etching mode when there is no surface structure, but there exists a bulk structure, which is confirmed by the variation of the volt-ampere characteristics of the tunnel current (figure 3(b)). The availability of such structure at the surface involves difficulties for development of adhesive diffusion layer at the metal–coating boundary that is non-desirable during VIP treatment. Therefore the time of effective influence of the gas plasma on the surface can be estimated according to the change in energy state of the surface when the contaminating adsorbed layers and oxides are removed and the V_p value transits from the negative domain to the positive one.

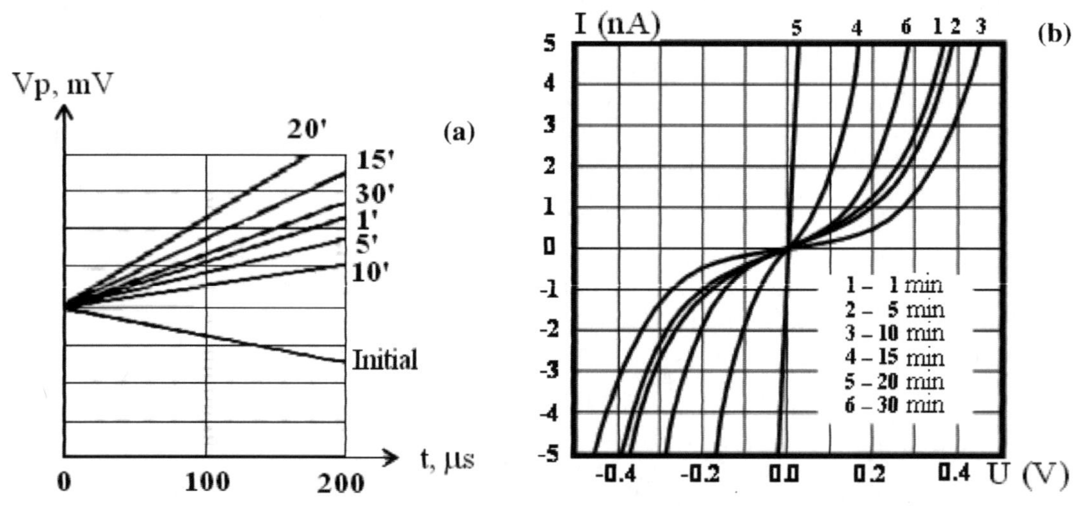

Figure 3. Change in the properties of the surface layer of steel during ion etching (U = –500V) depending on the time of argon gas plasma stream action: change in the energy state (a); variation in the volt-ampere characteristics of the tunnel current (b).

Tooling effects on the surface of structural metal materials lead to the structural changes of the surface layer, generation of a relief with different roughness and change in its energy state.

Thus the method set forth in this paper for measuring variations in values of the energy state of the surface layer after it was affected by plasma streams enables us to estimate their effectiveness and substantiate modes of their implementation.

References

[1] Sulima A I, Shulov V A and Yagodkin Yu D 1988 *The Surface Layer and Performance Properties of Machine Elements* (Moscow: Machinostroenye)

[2] Bobrov G V and Iliyin A A 2004 *Application of Inorganic Coatings* (Moscow: Internet Engineering)
[3] Iliyin A A, Plikhunov V V and Petrov L M 2011 *Aviation Industry* **2** 28–32
[4] Mush V N, Plikhunov V V and Petrov L M Patent RU 2471198
[5] Plikhunov V V and Petrov L M 2011 *Proceedings of the 10th International Conference "Films and Coatings"* pp 147–50

Journal of Physics: Conference Series **729** (2016) 012012 doi:10.1088/1742-6596/729/1/012012

Magnetron sputtering in rigid optical solar reflectors production

O Kh Asainov, D D Bainov, V P Krivobokov and D V Sidelev

Tomsk Polytechnic University, Tomsk, Russia

E-mail: das@tpu.ru

Abstract. Magnetron sputtering was applied to meet the growing need for glass optical solar reflectors. This plasma method provided more uniform deposition of the silver based coating on glass substrates resulted in decrease of defective reflectors fraction down to 5 %. For instance, such parameter of resistive evaporation was of 30 %. Silver film adhesion to glass substrate was enhanced with indium tin oxide sublayer. Sunlight absorption coefficient of these rigid reflectors was 0.081–0.083.

Optical solar reflector (OSR) is an important part of thermal control system of satellites. This element consists at least of thin glass substrate which back side is covered with silver based multilayer coating [1]. Glass substrate has low coefficient of sunlight absorption (A_s) and so it does not heat under influence of solar radiation. The substrate thickness is usually of 0.1 mm. It is enough for protection of reflective coating and provision of lightweight of the substrate.

Reflective coating includes a few layers: silver film with thickness of 110–130 nm, protective layer (20–30 nm) and adhesive layer (10–20 nm). The first one usually is deposited directly on the substrate. Protective Ni80Cr20 layer prevents damages of reflective film during storage of the element or it mounting on a satellite. Adhesive silicon oxide layer is used when the reflector is fixed on spacecraft surface by means of conductive silicate glue.

The growing need for glass OSRs demanded to raise productivity of reflective coating deposition process as well. Resistive evaporation in vacuum provides high reflection coefficient of silver layer. In terms of this method, productivity growth could be reached by mounting the greater number of evaporators in working chamber in order to process more quantity of glass substrates within the coating deposition cycle. But in this case the uniformity of deposited layers thickness goes down and it causes rising of defective reflectors fraction. The reduction of this negative effect by adjusting the working mode and relative position of evaporators, using special designed evaporators, etc., is still a quite difficult task.

There are many methods of thin film deposition and among them the magnetron sputtering provides efficient combination of deposition process productivity and coatings quality, especially when flat large substrates need to be covered with a thin film. Operating principle and design of magnetron sputtering system (MSS) enable to get thin layers with non-uniformity in thickness less than ±5 % along the length of this plasma source. In case of conductive targets the coating growth rate is a few tenths to several nm/s for oxides and about ten to a few dozen nm/s for metal films. The length of MSS can be relatively easy increased up to 2.5 m without loss of homogeneity in thickness of deposited coating and therefore large flat substrates are treated uniformly [2].

The laboratory plant was used to carry out magnetron sputtering deposition of reflective coating on glass substrates. The plant was equipped with mobile 500×500 mm^2 substrate holder which was able to move back-and-forth under fixed 700×100 mm^2 rectangular magnetron sputtering system. Distance between its target and holder surface was 100 mm. Power supply of magnetron sputtering system was provided by MF (130 kHz) DC unit.

There were some doubts about reflection coefficient of the silver layer formed by magnetron sputtering, but result of experiments showed $A_s = 0.08$ for the OSR with such layer. This value met to requirements and it was little bit more then in case of resistive evaporation ($A_s = 0.07$).

It was decided to apply the ion beam (3 kV, 0.05 A, Ar^+) for the surface treatment before silver layer deposition. This process took 0.5–1 h and it was enough to enable the layer passed the type test. The ion beam was generated by closed-drift ion source [3] which length was the same as that of the magnetron sputtering system.

Magnetron sputtering deposition of reflective layer on 100 glass plates (40×40 mm^2) placed on substrate holder showed that defective OSRs fraction decreased down to 5 %. For instance, such parameter of resistive evaporation was of 30 %. This fraction was estimated on the base of information about uniformity of A_s and coating thickness all over the surface of samples chosen from different positions on holder. The investigation of A_s was carried out by photometer FM-56. Thickness was measured by microscope-interferometer MII-4.

Results of the experimental work became one of the main reasons to design the vacuum plasma plant for rigid OSR making [4]. The plant was equipped with one ion source and four magnetron sputtering systems. Three of them were purposed to deposit reflective coating and one had indium(95 %)-tin(5 %) alloy target for obtaining indium tin oxide (ITO) layer on face side of the OSR. This transparent conductive oxide layer prevents OSR front surface charging [1].

During the technological testing of the plant there was an adhesion problem. After deposition it was found that the reflective coating held on substrate very good, but within a few days its adhesion became low and so the coating was not able to pass the type test. This difficulty most likely was caused by internal tension of silver layer. Some measures such as variations of working regime of the magnetron sputtering system with silver target, application of Ni80Cr20 film as an adhesive sublayer, etc., were taken but they were unsuccessful. In the latter case, A_s grew above upper limit of 0.086 as well, even when this film was deposited with minimal discharge power.

Under the circumstances only one solution of the problem was ITO adhesive sublayer (figure 1). Thickness of this film was chosen as thin as possible to provide lowest influence on A_s and required adhesion. In this way reflective coating adhesion was improved and stabilized while A_s of the OSRs was 0.081–0.083. The plant testing was successfully completed and it was accepted to produce rigid OSRs for satellites.

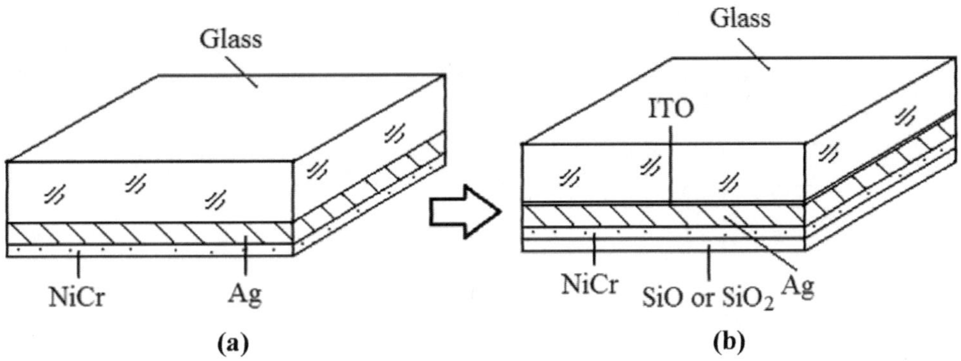

Figure 1. Evolution of the rigid OSR structure: (a) – basic composition; (b) – design with adhesive ITO sublayer and SiO or SiO_2 film.

Journal of Physics: Conference Series **729** (2016) 012012 doi:10.1088/1742-6596/729/1/012012

Thus, plasma magnetron sputtering enabled to get silver based reflective coating with good quality and uniformity of thickness on greater number of glass substrates. This resulted in significant decreasing of defective OSRs fraction and productivity of thin film deposition process growing. But technology on the base of the plasma method has to be developed taking into account its advantages and drawbacks, for instance internal tension of silver layers obtained by magnetron sputtering.

References

[1] Boshma J 1982 *Proceedings of an International Symposium on Spacecraft Materials in Space Environment (Toulouse)* pp 159–73

[2] Martin P M 2010 *Handbook of Deposition Technologies for Films and Coatings* (Oxford: Elesvier)

[3] Anders A 2005 *Surface and Coatings Technology* **5–6** 1893–906

[4] Ermolaev R A, Yevkin I V, Mironovich V V, Kharlamov V A, Khalimanovich V I, Asainov O H, Krivobokov V P 2009 *Perspective Materials, Devices and Structures for Space Applications: The Thematic Workshop (Yerevan)* (Yerevan: YerPhI) pp 54–7

Journal of Physics: Conference Series **729** (2016) 012013 doi:10.1088/1742-6596/729/1/012013

Thin-film heterostructures based on oxides of copper and zinc obtained by RF magnetron sputtering in one vacuum cycle

V Afanasjev, M Bazhan, B Klimenkov, N Mukhin and D Chigirev

St. Petersburg State Electrotechnical University, ul. Prof. Popova 5, St. Petersburg, 197376, Russia

E-mail: VPAfanasjev@mail.ru, muhinnv_leti@mail.ru

Abstract. Investigations of formation conditions of oxide heterostructures ZnO/CuO in the same vacuum cycle using RF magnetron sputtering of powder targets of zinc and copper oxides were carried out. The optical and electrical properties of the thin film structures were studied.

1. Introduction

Among new promising photoactive materials some oxide semiconductors such as Cu_2O, CuO, ZnO and heterostructures on their basis take a special place. The advantages of oxide systems are their environmental friendliness during manufacturing, operation and disposal. However, the present problem of use these materials for photovoltaic solar cells is characterized by low efficiency and irreproducibility experimental results.

CuO and *p*-type Cu_2O oxide semiconductors have great potential in use as an active layer of the solar cell due to the fact that these materials are of low cost, are widely distributed in the world, have a bandgap in the range of 1.21...2.1 eV [1], relatively high coefficient of absorption spectrum of sunlight. For Cu_2O the theoretical estimation of converting efficiency of solar energy into electrical energy is 9…12 %.

ZnO as an *n*-type inorganic semiconductor, is widely used in photovoltaics as an electron collection and hole blocking material due to its salient characteristics such as good optoelectronic properties, easy synthesis, non-toxicity. It has optical band gap at room temperature in the range of 3.1...3.3 eV [2].

For ZnO/CuO heterojunctions variety of methods including solid state synthesis, pressing, co-precipitation, sol-gel method, magnetron sputtering, thermal oxidation and hydrothermal method were applied [3–7]. For solar cells based on Cu_2O maximum efficiency of ~ 3.8 % was obtained in [8], where authors used high-temperature annealing and pulsed laser deposition. Among the main ways of increasing the efficiency of solar cells based on CuO/ZnO and Cu_2O/ZnO, the following can be identified: improving quality of material, direct doping can improve electrical properties of the oxides of copper and zinc to accumulate charge carriers; improving uniformity of layers can minimize leakage current and increase fill factor.

The aim of the work was study of the processes of formation of thin film heterostructures on the basis of copper and zinc oxides by high frequency magnetron sputtering of powder targets.

2. Obtaining of the structures

Thin film ZnO/CuO heterostructures were prepared by applying zinc oxide and copper by reactive high frequency magnetron sputtering on a glass substrate. The targets for the deposition of zinc oxide

Journal of Physics: Conference Series **729** (2016) 012013 doi:10.1088/1742-6596/729/1/012013

films and copper oxide powders used chemically pure ZnO and analytical grade CuO. Thin films of copper were deposited by magnetron sputtering at room temperature under an argon atmosphere at a pressure of $2\cdot10^{-5}$ bar. The substrates were placed at a distance of about 20 cm from the target. In order to control the optical properties of ZnO samples near fastened clean glass substrates. After the deposition, films were annealed in air to form a layer of copper oxide and ZnO–CuO structures. Annealing was carried out either in a muffle furnace or in a quartz furnace pulse thermal annealing.

The performed set of experiments showed that the studied oxides are characterized by low spraying speeds; were obtained relations between the growth velocity of zinc and copper oxide and the power of high frequency discharge and worked out technological spraying regimes of layers. Thin film samples of oxide structures were fabricated, which represented the following sequence of layers: glass/*n*-ZnO/*p*-CuO/Me and glass/*n*-ZnO/ZnO/*p*-CuO/Me (figure 1). As electrode materials (Me) were used Cu, Pt, Ag, Ni, Al. The best results were obtained with Pt electrodes.

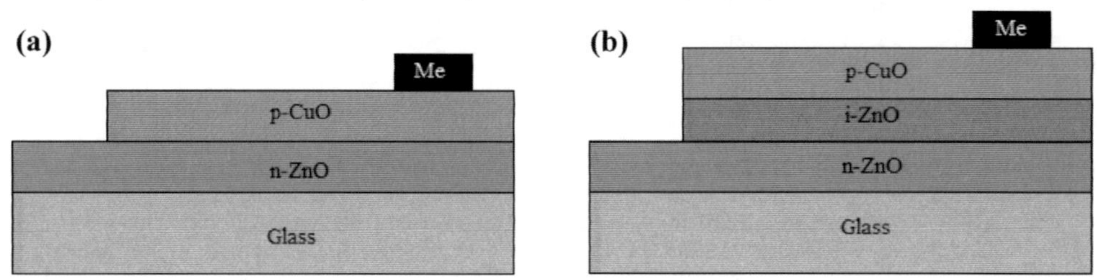

Figure 1. Schematic representation of obtained structures: (a) – *p*-CuO/*n*-ZnO; (b) – *p*-CuO/*i*-ZnO/*n*-ZnO.

3. Study of the obtained structures

Investigation of electrophysical parameters of copper oxide films was carried out on the installation ECOPIA HMS-5000. Measurements showed that samples have *p*-type conductivity and have a high surface resistance. Also the transmittance and the reflection spectra of the films were measured. Measurements were performed on AvaSpec-2048 spectrophotometer in the visible range of 400...1000 nm using a fiber-optic module for measuring reflectance spectra. The estimations of the band gap of studied samples was 1.6...1.7 eV. Optical studies have shown that while reactive high frequency magnetron sputtering of copper oxide powder target took place, copper oxide (II) is deposited on a substrate, which is also confirmed by X-ray phase analysis data.

The reflection and the absorption spectra of CuO films showed no thermal effect up to 350 °C inclusive, that indicating stability of the formed oxide. Zinc oxide films had transmittance of about 80 % in the wavelength range from 400 to 1100 nm.

Studies of current-voltage characteristics of oxide ZnO/CuO heterostructures have been carried out in dark mode and under a light source with a spectral response close to the solar spectrum. The structures with additional layer of ZnO shows better I–V characteristics (figure 2).

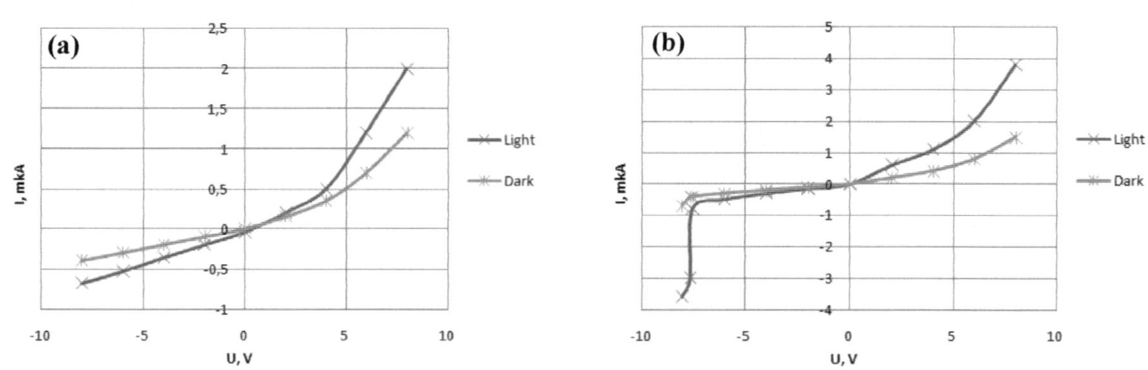

Figure 2. I–V characteristics of *p*-CuO/*n*-ZnO (a) and *p*-CuO/*i*-ZnO/*n*-ZnO (b).

This difference is explained by the fact that the layer of ZnO align energy characteristics in the band diagram (figure 3). Experimental data on the characteristics of the electron transport in thin-film oxide heterojunction structures with CuO/ZnO were obtained.

It was found that in the samples, where copper oxide was deposited at elevated temperatures up to 300 °C, characteristics differed by weak nonlinearity and of existence of a large series resistance, that was presumably a consequence of the degradation of conductive zinc oxide layer in the structure.

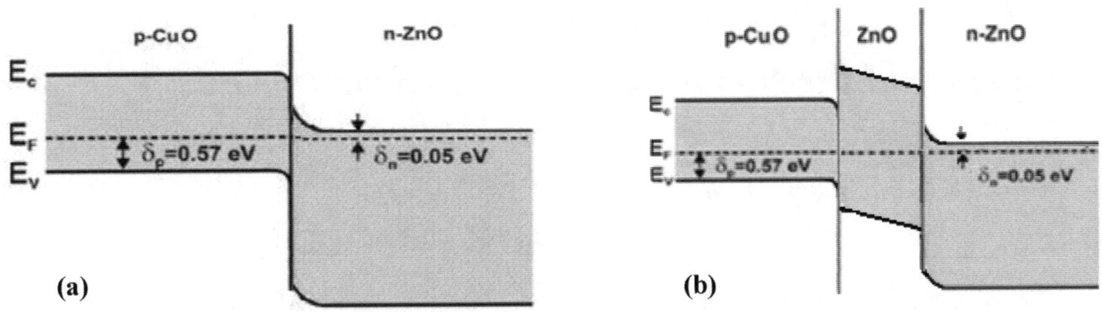

Figure 3. Energy band structure in equilibrium: (a) – *p*-CuO/*n*-ZnO; (b) – *p*-CuO/*i*-ZnO/*n*-ZnO.

The samples, that were formed at low temperatures, also showed weak photoresponse and high series resistance, presumably because of not well-formed crystal structure of the films. Selection of medium temperatures (about 200...250 °C) allowed to reduce the series resistance of the structure, and adding a part of the structure of heterojunction *n*-ZnO/*p*-CuO thin intermediate layer of zinc oxide (ZnO) can improve the nonlinearity of the current-voltage characteristics and improve the photoresponse heterostructures. However, the results need further optimization.

4. Conclusions

Thus, in this work the high frequency magnetron sputtering of zinc and copper oxide powder target was first used to form heterostructures ZnO/CuO in one vacuum cycle. Research of electrophysical and optical parameters of obtained structures was made. There is a need in improving in the technological parameters in the manufacturing process to use similar structures in photovoltaics.

Acknowledgments

This work was financially supported by the grant of the Russian Science Foundation (project № 14-12-00327).

References

[1] Mugwang'a F K, Karimi P K, Njoroge W K *et al* 2013 *J. Thin Film Sci. Tec.* **2** 15–24
[2] Srikant V and Clarke D R 1998 *J. Appl. Phys.* **83** 5447
[3] Aygün S and Cann D 2005 *Sensors and Actuators B* **106** 837–42
[4] Saji K J, Populoh S, Tiwari A N and Romanyuk Y E 2013 *Physica Status Solidi A* **210** 1386–91
[5] Nagaraju G, Ko Y H and Yu J S 2014 *Materials Letters* **116** 64–7
[6] Omayio E O, Karimi P M, Njoroge W K and Mugwanga F K 2013 *Int. J. Thin Film Sci. Tec.* **1** 25–8
[7] Fujimoto K, Oku T, Akiyama T and Suzuki A 2013 *Journal of Physics: Conference Series* **433** 012024
[8] Minami T, Nishi Y, Miyata T and Nomoto J 2011 *Applied Physics Express* **4** 062301

The role of current characteristics of the arc evaporator in formation of the surface metal-coating composite

V V Plikhunov[1], L M Petrov[1] and K V Grigorovich[2]

[1] Joint Stock Company "National Institute of Aviation Technologies", Moscow, Russia
[2] Federal State Budgetary Scientific Institution "Institute of Metallurgy and Materials Science after A.A. Baykov", Russian Academy of Sciences, Moscow, Russia

E-mail: plm@niat.ru, grigorovichkv@gmail.com

Abstract. The influence of current characteristics of the vacuum arc evaporator on the interaction process of plasma streams with the surface under treatment during generation of the physicochemical properties of the formed metal-coating composite is considered. It is shown that the interaction of plasma streams with the processed surface provides surface heating, defects elimination, change in energy properties, and mass transfer of plasma stream elements activating surface diffusion processes whose intensity is evaluated by the arc current magnitude and location of the processed surface relative to the cathode axis.

1. Introduction

Plasma streams affecting the surface form the processes, the effectiveness of which is determined by the energy of components of elemental composition of gas-metal plasma streams. The result of such effects are the processes associated with surface layer heating, layerwise controlled removal of a defective absorbent-oxide layer, deposition of plasma stream elements, surface plasma-chemical reactions, interdiffusion saturation of the surface layer with elements of the plasma stream and the substrate [1–3].

The sequence of plasma stream effects on the surface being processed and their interaction generate technological stages of the surface treatment processes. Development of a wide range of external multilayer, multicomponent, gradient, nanostructured coatings is accompanied by changes in the chemical composition of the surface layer, chemical activity, structural and phase composition of the surface. Generation of the deposited coating properties depends on many factors including components ratio of gas-metal plasma stream determined by the current characteristics of the arc evaporator. This is especially urgent in formation of coatings where the system of obtained properties depends upon the change in the ratio of plasma chemical reaction components achieved through the variation in arc current of the evaporator and parameters of gaseous medium pressure.

2. Materials and experimental procedure

Model specimens made of 40H steel (150×30×2 mm) produced to manufacturing technology were investigated. Effects of titanium–nitrogen plasma stream on the surface under investigation were realized during vacuum ion-plasma treatment in a universal vacuum ion-plasma unit RADUGA developed and manufactured by JSC NIAT. The deposition process of titanium plasma stream occurred when one evaporator operated in nitrogen atmosphere with constant values of residual pressure ($P = 3\cdot10^{-1}$ Pa), reference voltage value ($U_{ref} = -100$ V) and time of plasma stream effect (30 min). The variable value was the arc current ($J_{arc} = 60$, 100, 150 A). Titanium plasma deposition was preceded by electron heating to the temperature of 360 °C and ion etching of the surface under treatment by argon plasma ($U_{ref} = -500$ V, 3 min).

Investigation of the initial parameters of the processed surface layer and evaluation of the properties variation after manufacturing effects as well as interactions with the titanium–nitrogen plasma stream were conducted using the following devices and equipment: change of the element chemical composition was carried out with Alpha–2000 class V portable metal and alloy analyzer made by Innox-X Systems company; evaluation of the change in the energy state of the surface layer was done by contact potential difference method according to the change of the value of reduced surface potential V_p (mV); surface layer hardness was investigated using a MicroMet 5101 device with loading from 0.1 to 2 N to measure surface hardness; layer-by-layer analysis of the quantitative change in the chemical composition of the surface layer was performed by a glow discharge atomic emission spectrometer GDS 850A (LECO). The depth of the layer-by-layer analysis measurement range was 0.01 to 25 μm.

3. Experimental results and discussion

The process of metal plasma stream generation is fundamental in formation of any type of coatings using the vacuum ion-plasma processing method. Separation of plasma stream generated by the electric arc evaporator into electron-ion components makes it possible to enhance significantly the capabilities of manufacturing effects of the streams during processing. Impact of the electron flow (anode – processed part connection scheme) has allowed for improving effectiveness of the stage of processed surface heating, which consists in its uniformity and speed, herewith the maximum value is obtained at the end of the heating stage, figure 1.

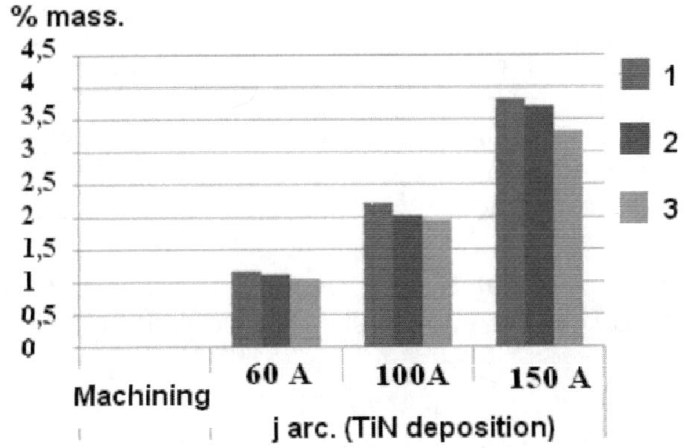

Figure 1. Effect of the arc current on Ti mass transfer along the specimen length: 1 – specimen area 70 mm above the center; 2 – central area of the specimen; 3 – specimen area 70 mm lower than the center.

Temperature elevation of the processed surface is determined by its kinetic interaction with the electron component of plasma stream and transferred in that case by electron release energy. Subsequent effects of plasma stream on the surface being processed change its heating temperature since the value of reference current obtains prevalent significance. Transfer to the stage of ion etching in conditions of the semi-self-maintained gas discharge leads to temperature lowering in the surface layer at the cost of reduction in plasma stream energy caused by decrease of the reference voltage value. Realization of the deposition stage of plasma stream elements is associated with decrease in the reference voltage value that makes for further decrease with the following stabilization of the surface temperature. Variation of the current values of the evaporator in the range of 65 to 150 A actually does not influence the stage behavior of the change in the temperature dependence upon realization of the manufacturing effect stages.

Temperature of the surface layer and the presence of primary energy of plasma stream ions at the stage of deposition promote mass transfer of plasma stream elements on the substrate under treatment and interdiffusion of elements both of the substrate and the plasma stream, which are determining factors for the composite formation process. Variation of the arc current within the range of 60 to 150 A with all the rest process parameters being constant considerably affected the character of change in properties of the formed surface layer. This is due to the fact that the increase of the arc current enhances the density of plasma stream in the vacuum chamber volume and consequently increases the mass of the deposited element (Ti) in the coating, figure 1. With rise in arc current the non-uniformity of mass distribution of the deposited plasma elements along the specimen length also increases. The presence of gas-metal components (Ti–N) in the plasma stream results in their interaction with the processed surface and occurrence of plasma chemical reaction processes at the surface to form Ti–N system compounds.

According to the state diagram (Ti–N) the given system possesses a large domain of existence of titanium–nitrogen compounds with a variable hardness value determined by the ratio of plasma stream components (titanium and nitrogen) entering into a reaction. Dissolving in the titanium the nitrogen forms interstitial solid solutions, which transit into chemical compounds as the nitrogen concentration increases. Growth of the arc current value promotes change in the correlation of plasma stream components (titanium and nitrogen) as well as increases the mass transfer of titanium, which in turn increases the thickness of the coating under formation. An increase in hardness values in relation to the arc current, Figure 2, most likely depends on the factors that are determined by the growth of thickness of the generated coating rather than the change in the correlation of the stream components.

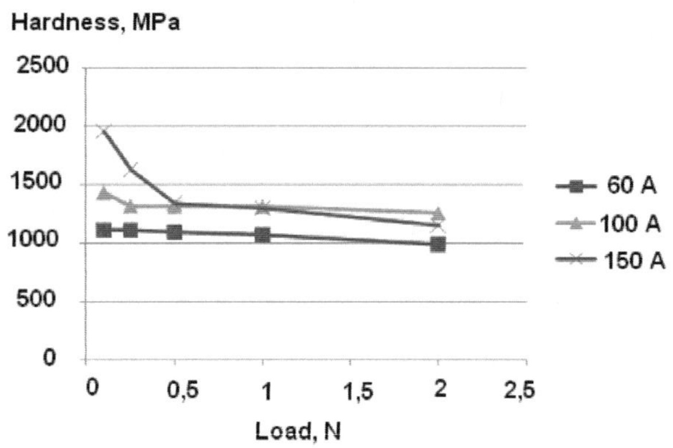

Figure 2. Variation of hardness in the surface layer of the 40H alloy formed during interaction with the Ti–N system plasma stream in relation to the arc current.

The change in the energy state of the surface evaluated by value of the reduced surface potential (V_p, mV) has demonstrated a weak dependence on the arc current value. It is connected with formation of identical structural state in the surface layer, figure 3.

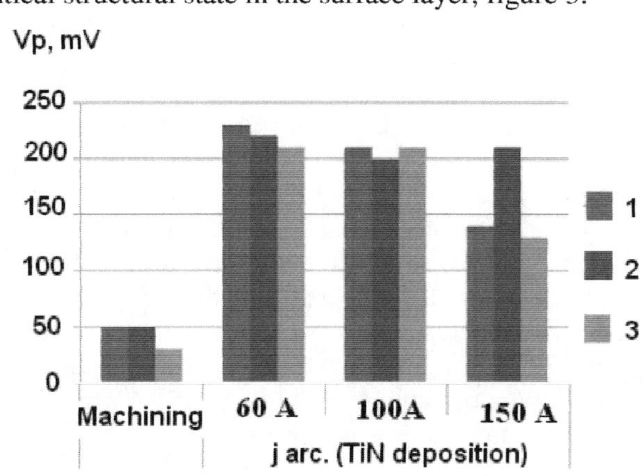

Figure 3. Effect of the arc current on variation of the reduced surface potential V_p of the 40H steel along the specimen length: 1 – specimen area 70 mm above the center; 2 – central area of the specimen; 3 – specimen area 70 mm lower than the center.

Layer-by-layer analysis that records changes of the element composition in the surface layer formed, when affected by the TiN system plasma stream on the 40H steel surface, has shown that arc current significantly influences diffusion processes along with correlation of plasma stream elements entering into a plasma chemical reaction that forms coatings, figure 4. Deposition of the TiN functional coating is associated with the increase of the Ti mass transfer on the substrate, especially in the area along the cathode axis. The most intense increase of Ti along the entire length of the specimen is observed at the high value of arc current (150 A).

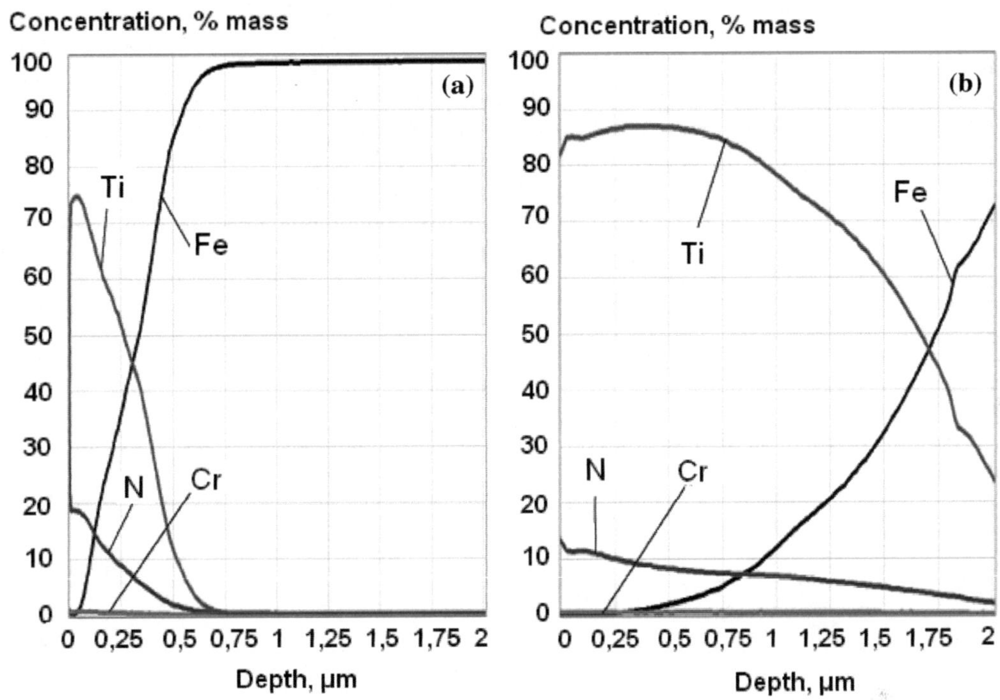

Figure 4. Distribution of the surface layer elements of the 40H steel formed under the influence of the Ti–N system plasma stream in relation to the arc current: (a) – J_{arc} = 60 A; (b) – J_{arc} = 150 A (processed surface is 70 mm higher than the cathode axis).

Clearly traced is a tendency for an increase in the current; titanium–nitrogen diffusion zone grows into the surface layer of the steel specimen. Availability of such a diffusion zone is a guarantee for adhesion strength of the formed coating; the larger is the diffusion zone, the higher is the adhesion strength of the coating to the substrate.

4. Conclusions

Variation in the arc current in the range of 60 to 150A caused titanium mass transfer to increase 3.5 times with a moderate spread in values along the specimen length. The resultant dependence of the change in mass transfer value of the Ti–N system plasma stream elements on the arc current provided an increase in thickness of the formed coatings and improvement in values of surface hardness during their interaction with the surface being processed.

Energy state of the surface layer evaluated by values of the reduced surface potential demonstrated that the coating, which was formed with different current arc characteristics, had virtually a constant value of V_p with minor deviations. Such character of V_p value variation is indicative of the similar structural state at the surface.

Layer-by-layer analysis that records changes of the element composition in the surface layer formed, when affected by the TiN system plasma stream on the 40H steel surface, has shown that the arc current significantly influences diffusion processes along with correlation of plasma stream elements entering into a plasma chemical reaction that forms coatings. At the top part of the area under investigation (70 mm above the cathode axis) we observe correlation dependences of the coating thickness growth, the depth of interdiffusion of elements of both the substrate and plasma on values of the arc current. To the maximum arc current corresponds the maximum titanium mass transfer and the maximum effectiveness of diffusion processes. Along the cathode axis and in the lower part of the area under investigation (70 mm below the cathode axis) correlation dependences established above completely remain, however, effectiveness of their interaction with the substrate is slightly reduced (coating thickness, depth of interdiffusion of substrate and plasma elements).

References

[1] Ilyin A A, Plikhunov V V, Petrov L M, Ivanchuk S B and Gavrilov A S 2006 *Congress abstracts of 5th International Congress IAC'06 (Moscow)* pp 160–1

[2] Plikhunov V V, Petrov L M, Ivanchuk S B and Gavrilov A S 2008 *Proceedings of the VII scientific conference on hydroaviation "Hydro-Air Show-2008" (Gelendzhik)* pp 233–8

[3] Plikhunov V V and Petrov L M 2010 *Proceedings of the IX scientific conference on hydroaviation "Hydro-Air Show-2010" (Gelendzhik)* pp 206–11

Journal of Physics: Conference Series **729** (2016) 012015 doi:10.1088/1742-6596/729/1/012015

An influence of technological parameters of plasma-chemical deposition of SiO_2 films on their electro-physical properties

A A Romanov, A V Serkov and E S Hruleva

St. Petersburg Electrotechnical University "LETI", Professora Popova str. 5, St. Petersburg, 197376, Russia

E-mail: cmid_leti@mail.ru

Abstract. In this article a formation process of dielectric films on silicon (100) and silicon carbide using plasma-chemical deposition is described. Experimental relationships of SiO_2 films thickness and main technological parameters are presented. Values of electro-physical parameters of films are measured.

1. Introduction

Formation of high quality dielectric coatings is an increasingly important problem, which needs an efficient solution. Such coatings are widely used in microelectronics as:

- protecting and passivating layers on surfaces of products;
- masking layers for selective plasma and liquid etching;
- layers for side-wall local isolation during formation of gate block of MOS-transistors.

In modern microelectronics most widely used are thin films of SiO_2 and Si_3N_4. In this work the objects for testing are films of amorphous SiO_2, deposited onto (100)-Si and 6H-SiC substrates using method of plasma-chemical deposition in a mixture of vapors of oxygen and hexamethyldisiloxane (HMDS) with the help of apparatus URM-053 (Russia) (figure 1).

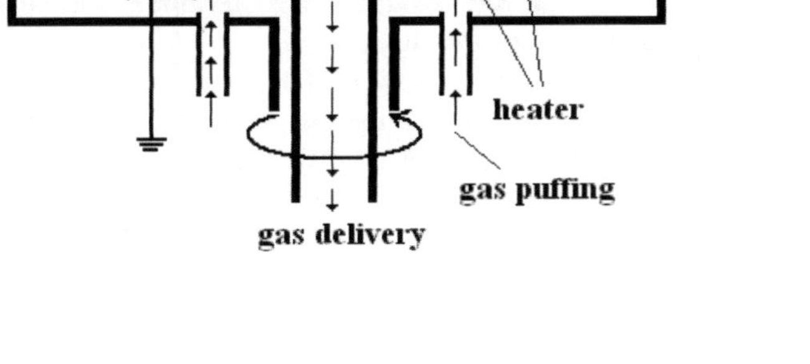

Figure 1. Schematic representation of reaction chamber of apparatus URM-053 (Russia) for deposition of dielectric films using PECVD.

In PECVD technology for decomposition of reaction gas into active radicals gaseous plasma is used. Thermodynamic non-equilibrium of decomposition processes in gas discharge enables deposition of amorphous and polycrystalline films to be carried out at considerably lower temperatures, in comparison with analogical processes of gas phase chemical deposition (CVD) with thermal decomposition of reaction gas.

In PECVD process chemical activation is achieved due to decomposition of gas mixture into electrons, ions and active radicals. Radicals and ions, approaching substrate enter into a reaction in its surface yielding amorphous or polycrystalline layer of target material.

The main stages of PECVD process are:
– vaporization of initial reagent;
– chemical activation of gas molecules;
– transport of reagent to substrate using argon as a carrier-gas;
– plasma stimulated reaction yielding the solid product;
– removal of gaseous reaction products and surplus reagent from the reaction zone (figure 2).

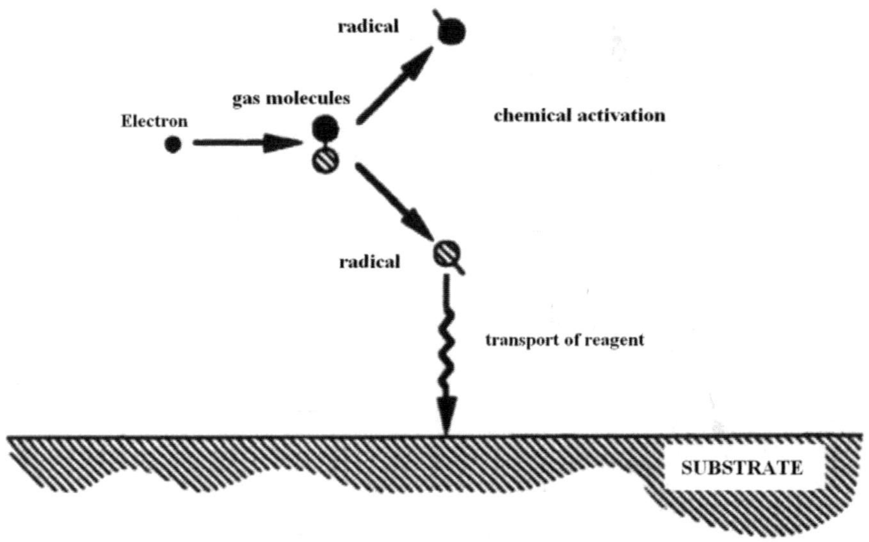

Figure 2. Main stages of PECVD process.

2. Experimental

Three-stage process being used for plasma-chemical deposition of SiO_2 onto (100)-Si and 6H-SiC substrates. At the first stage substrates were treated in oxygen for preparation of natural SiO_2. At the second stage hexamethyldisilane (HMDS) was fed into chamber where deposition of "main" SiO_2 layer was carried out. At the third stage substrate with deposited layer was treated in oxygen, for SiO_2 structure contraction and prevention of defect formation in bulk SiO_2.
Chemical reaction on the surface of the substrate using HMDS could be presented as follows:

$$Si_2O(CH_3)_6 + 12O_2 \xrightarrow{200\text{–}300\ ^{\circ}C} 2SiO_2\downarrow + 9H_2O\uparrow + 6CO_2\uparrow.$$

Properties and composition of layer being prepared is defined by ratio gas mixture components, discharge power, substrate temperature and conditions of ion bombardment.

The following technological parameters were variated: W – power of HF-generator of plasma; t_1, t_2 – treatment time in HMDS and oxygen plasma, correspondingly. Delivery rate ratio of gas mixture O_2/HDMS (5:1) and treatment time in oxygen at the first stage (t = 10 min) remained constant for all experiments. Refractive index and thickness of the film were measured with the help of ellipsometer LEF-3M (Russia) (figures 3, 4). Refractive index was of weak dependence of technological parameters for all films and was $n \sim 1.46$.

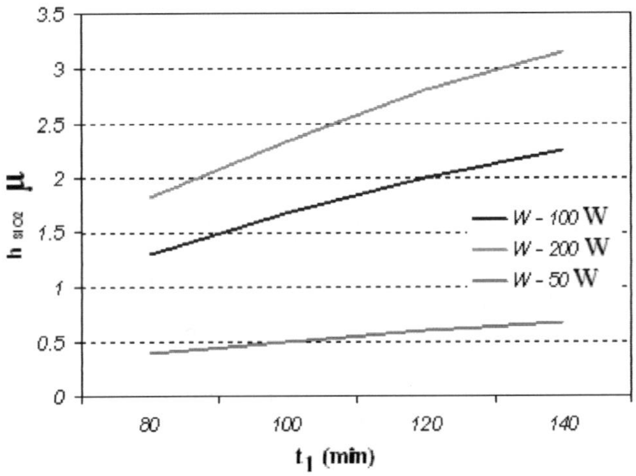

Figure 3. Relationships of SiO_2 film thickness vs. treatment time in HDMS (t_1) at different values of HF-generator power and equal treatment time in oxygen at the third stage (t_2 = 10 min).

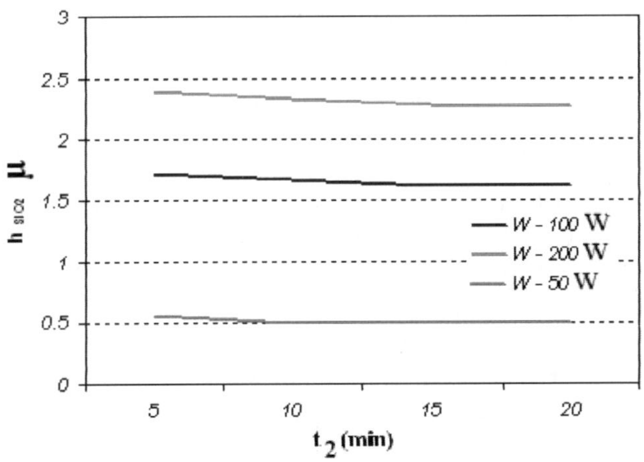

Figure 4. Relationships of SiO_2 film thickness vs. treatment time in oxygen at the third stage (t_2) under different values of HF-generator power and equal time of HDMS treatment (t_1 = 100 min).

From figure 3 one can see that film thickness grows with the increase of HDMS treatment time as well as with increase of power, because in the latter case amount of chemically active particles in plasma grows. In figure 4 decrease of film thickness is connected with material contraction after treatment in oxygen plasma.

Structures of the films were investigated using FIB-complex «Helios Nanolab». It has been shown that under HF-generator power of 200 W, HDMS treatment time of 120 min and oxygen plasma treatment time at the third stage of 10 min, the density of layers was ~ 2.27 g/cm^3.

Main electro-physical characteristics of amorphous SiO_2 layers were measured using probe station – immitance measuring device E7-20 (Russia), namely:

1. Specific electrical resistance strongly depends of layers purity and presence or absence of various defects;
2. Relative dielectric permeability at room temperature was 3…4;
3. Voltage proof was in the range 10^6…10^7 V/cm and depends on the concentration of defects in the film.

Journal of Physics: Conference Series **729** (2016) 012016 doi:10.1088/1742-6596/729/1/012016

Effect of structure and deposition technology on tribological properties of DLC coatings alloyed with VIA group metals

M Khrushchov[1], I Levin[1,2], E Marchenko[1], V Avdyukhina[2] and M Petrzhik[3]
[1] Institute of Machine Studies, Russian Academy of Sciences, Moscow, 101990, Russia
[2] Department of Physics, Moscow State University, Moscow, 119991, Russia
[3] National Research Technological University "MISiS", Moscow, 119049, Russia

E-mail: michel_x@mail.ru

Abstract. The results of a comprehensive research on atomic structure, phase composition, micromechanical and tribological characteristics of alloyed DLC coatings have been presented. The coatings have been deposited by reactive magnetron sputtering in acetylene-nitrogen gas mixtures of different compositions (*a*-C:H:Cr), by plasma-assisted chemical vapor deposition in atmospheres of silicone-organic precursor gases (*a*-C:H:Mo:Si), and by nonreactive magnetron sputtering of a composite target (*a*-C:H:W).

1. Introduction

Diamond-like carbon (DLC) coatings having amorphous carbon (*a*-C) structure are one of advanced materials for surface strengthening machine and mechanisms wear parts. They are characterized by high wear resistance and low friction even at high contact pressures. The active research is carried out on further improvement of wear resistance and reduction of the residual stresses in such coatings by alloying them with different metallic and nonmetallic elements.

The aim of the present work was to study the peculiarities of nanocomposite structure of plasma-assisted vapor deposited DLC coatings alloyed with chromium, molybdenum and tungsten and its relationship with deposition technology, as well as with mechanical and tribological characteristics.

2. Methods of coating characterization and synthesis

The investigations of fine atomic structure, phase composition, nanomechanical and tribological properties of DLC coatings obtained by plasma-assisted chemical vapor deposition from gaseous silicon-organic "precursors" (*a*-C:H:Mo:Si), by reactive magnetron sputtering in atmospheres containing acetylene (*a*-C:H:Cr) [1, 2] and by non-reactive magnetron sputtering of "mosaic" W/C targets (*a*-C:H:W) have been studied. The coatings deposited on steel substrates have been studied by X-ray diffractometry, energy-dispersive electron-probe microanalysis, instrumented nanoindentation. Raman spectroscopy has been used to estimate the state of carbon phase in coatings.

Tribological tests have been performed in conditions of dry friction using a one-ball "sphere-on-disc" tribometer at 0,02 to 0,2 N loads. Immovable silicon nitride ceramic spherical indentor (6 mm in diameter) has been used as counterpart.

Experimental X-ray spectra of coatings have been treated by the method described in [3] that made it possible to eliminate the contribution of the substrate in scattering intensity. To estimate the coating phase composition JCPDS PDF-2 and the "Match!" software have been used.

3. Results and discussion

3.1. Coating structure peculiarities

Table 1 and figures 1–3 represent the results of structural investigations. In *a*-C:H:Cr coating the formation of a complex structure formed by nanometer sized (~ 10 nm) chrome inclusions, carbide and nitride phases has been established (figure 1).

Table 1. Chemical and phase composition, nanohardness and structural state of coatings.

Coating type	Element concentration, at. %								Structure type	Phase composition	*H*, GPa
	C	N	O	Ar	Si	Mo	W	Cr			
a-C:H:Cr 100 % C_2H_2	80	4	–	–	–	–	–	15	nanocrystalline	$Cr_{23}C_6$ + Cr	10
a-C:H:Cr C_2H_2/N_2 = 20/80	27	29	–	–	–	–	–	43	nanocrystalline	Cr_3C_2 + Cr_2N + Cr	29
a-C:H:Mo:Si	40	–	8	–	36	16	–	–	amorphous	Mo_5Si_3 + Mo	15
a-C:H:W	75	–	2	0,9	–	–	17	–	amorphous	W + WC + W_2C	20

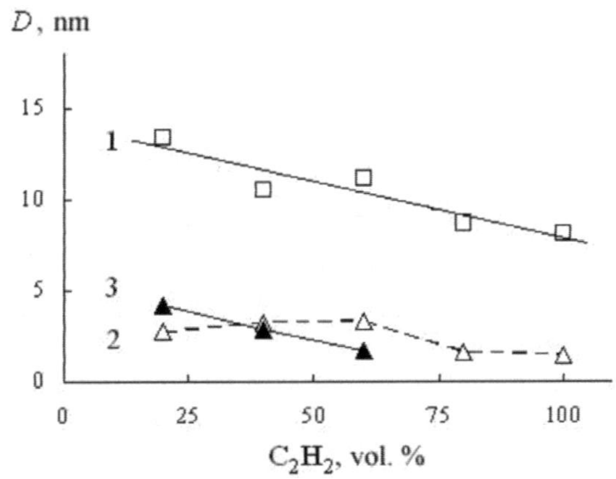

Figure 1. Coherently scattering domain sizes of chromium (1), its carbide (2) and nitride (3) phases in *a*-C:H:Cr coatings deposited in reactive gas mixtures with different acetylene content.

In coatings alloyed with molybdenum and tungsten the structure might be characterized as "X-ray amorphous". The proper diffuse scattering intensity for these coatings is presented in figures 2 and 3. Their modulations prove the absence of long-range order with conservation of short-range order elements at the distances of some nearest coordination spheres.

For the *a*-C:H:Mo:Si coatings this might be explained by the presence of ultra-dispersed inclusions of molybdenum, molybdenum carbide and silicide phases with size less than 0,5 nm the formation of which might be explained by the use of silicon-organic precursors during their synthesis. In DLC coatings alloyed with tungsten the structure might consist of metallicinclusions, tungsten carbides and *a*-C domains of sub-nanometer size.

These data demonstrate that the nature of nanocrystalline state in Me-DLC coatings may depend on the type of vacuum deposition technology used in their synthesis. So, one may conclude that reactive magnetron sputtering used for *a*-C:H:Cr deposition will result in a composite structure of nanosized domains of chromium and its interstitial phases that sustain the rather perfect long-range order in the matrix of amorphous carbon. As it has been shown in [2] the additional alloying of *a*-C:H:Cr with nitrogen leads to formation of chromium nitrides and is accompanied by the increase in nanomechanical and tribological properties.

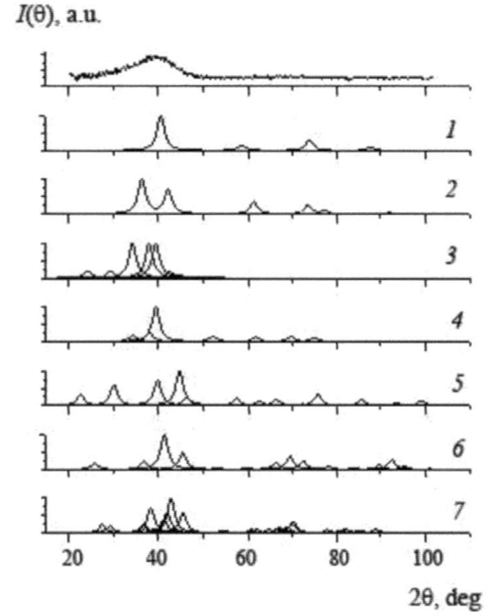

Figure 2. Experimental diffuse scattering intensity of *a*-C:H:Mo:Si coating (top) and the X-ray line positions for: *1* – Mo; *2* – MoC, *3* – Mo_2C (hex.); *4* – Mo_2C (orthorhomb.); *5* – $MoSi_2$; *6* – Mo_3Si; *7* – Mo_5Si_3.

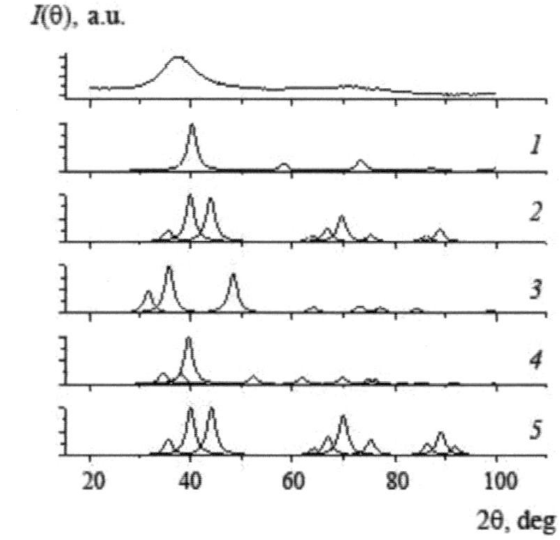

Figure 3. Experimental diffuse scattering intensity of *a*-C:H:W coating (top) and the X-ray line positions for: *1* – α-W; *2* – β-W; *3* – WC; *4* – W_2C; *5* – W_3C.

3.2. Tribological properties of coatings

The results of tribological tests are summarized on figure 4. The DLC coatings alloyed with chrome have demonstrated highest tribological parameters. For these coatings the friction coefficient *f* was lower than *f* = 0,2–0,35, and their complementary doping with nitrogen has allowed to make it lower, about *f* = 0,1 or less. At the same time the coating performance characterized by the number of cycles *N* before the fracture process initiation has reached its highest level for coatings deposited in pure acetylene and doping of coatings with nitrogen making *f* lower also at high contact pressures has been accompanied by a noticeable *N* decrease.

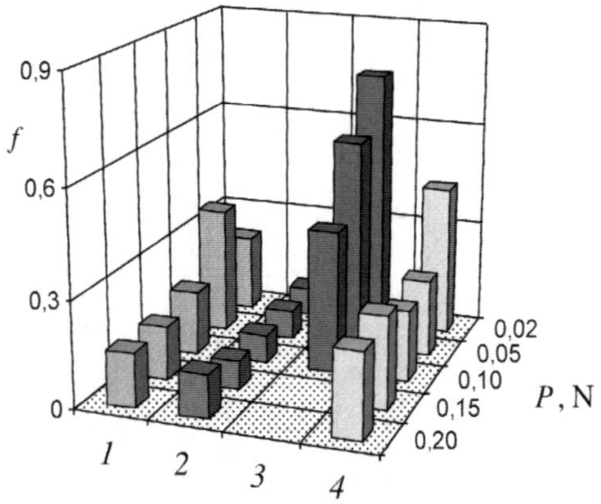

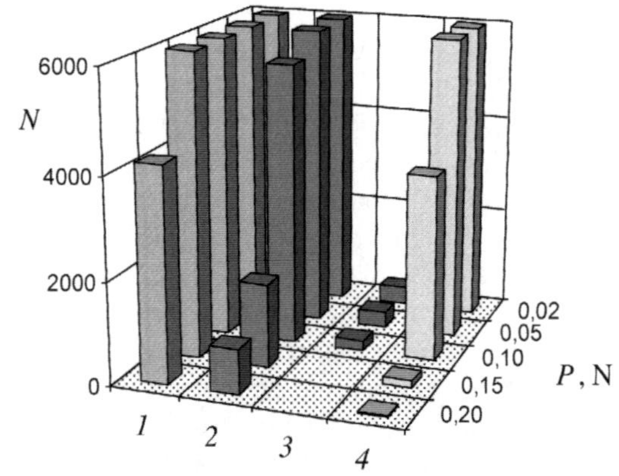

Figure 4. The dependence the coefficient of friction *f* (a) and the number of cycles *N* (b) before the coating fracture initiation on load for: 1 – *a*-C:H:Cr (deposited in pure acetylene); 2 – *a*-C:H:Cr (deposited in acetylene – nitrogen 20/80 gas mixture); 3 – a-C:H:Mo:Si; 4 – a-C:H:W coatings on KhN35VT substrates.

DLC coatings alloyed with W that have been deposited by non-reactive sputtering of a composite W/C target when tested have demonstrated $f \sim$ 0,2–0,4 for all the load values, that might be compared with f values for coatings deposited in pure acetylene but the results of their durability tests have been rather low.

The *a*-C:H:Mo:Si coatings as compared with the other ones have shown the highest f values and low durability. To our opinion this decrease of tribological parameters might be explained by the presence of molybdenum silicides $MoSi_x$ in them and/or by the processes of silicon hydroxides formation during friction in air environment. The presence of these hydroxides having low strength characteristics [1] leads to a coating mechanical strength decrease. The way to improve the tribological properties of these coatings might probably consist in lowering the contact pressures/changing contact geometry or using lubricants [1].

In general the results of the present investigation on the structure of DLC coatings alloyed with metals are in agreement with the known facts on the influence of the plasma-assisted deposition conditions on it. Thus the problem of nanocomposite structure formation in magnetron sputtered films of *a*-Cr-C has been discussed in [4] and the authors concluded that the structure with nanosized carbide inclusions is typical for reactive magnetron sputtering and the amorphous one – for a non-reactive technology.

3.3. Radial distribution function analysis

To our opinion to define the nature of phases formed in these conditions more precisely the analysis of the radial distribution functions $G(r)$ of coatings might be useful. These functions described the probably to find a pair of atoms at a distance r between them and the maxima of $G(r)$ correspond to atomic distances values being the most probable for a particular structure.

The "PDFgetX3" software [5] has been used to calculate $G(r)$ from the experimental X-ray diffuse scattering intensity $I(2\theta)$ spectra:

$$G(r) = \frac{2}{\pi} \int_{q_{\min}}^{q_{\max}} q\left[S(q) - 1\right]\sin(qr)dq, \quad (1)$$

where $q = 4\pi\sin\theta/\lambda$; θ – the diffraction angle; λ – the radiation wavelength; $q_{\min}$ and $q_{\max}$ correspond to $\theta_{\min}$ and $\theta_{\max}$ values respectively; $S(q)$ – Faber–Ziman type structure function:

$$S(q) = 1 + \left[I_{\text{coh}}(q) - \sum_i c_i \left|f_i(q)\right|^2\right] \cdot \left|\sum_i c_i f_i(q)\right|^{-2}, \quad (2)$$

c_i – atomic concentration; $f(q)_i$ – atomic factor of the i-th atom; $I_{\text{coh}}(q)$ – experimental diffuse scattering intensity.

The $G(r)$ functions for the coatings under study are presented on figure 5. One may notice that the doping element type and the deposition technology have a rather strong influence on the pair distribution function behavior and the local $G(r)$ maxima positions that might coincide with certain atomic pair distances characteristic for the crystal structures of the corresponding nanocomposite components. The radial distribution functions presented on figure 5 for $r \leq 10$ Å demonstrate strong oscillations that might decay at longer interatomic distances.

The analysis of the behavior of radial distribution functions $G(r)$ for Cr-DLC has demonstrated that in these coatings the oscillations of $G(r)$ decay only at $r >$ 100–120 Å (10–12 nm) that is close to the size of the coherently scattering domains, i. e. the elements of long-range order exist even for rather long interatomic distances. To our opinion this fact proves that the nanosized metallic and interstitial phase inclusions in chromium-based diamond-like nanocomposites retain the structure of their "parent" phases. For DLC coatings doped with tungsten and molybdenum (figures 5(c) and 5(d) respectively) the $G(r)$ oscillations decay at r about 8–10 Å and in their case only short-range correlations exist with a "correlation radius" not exceeding 1 nm.

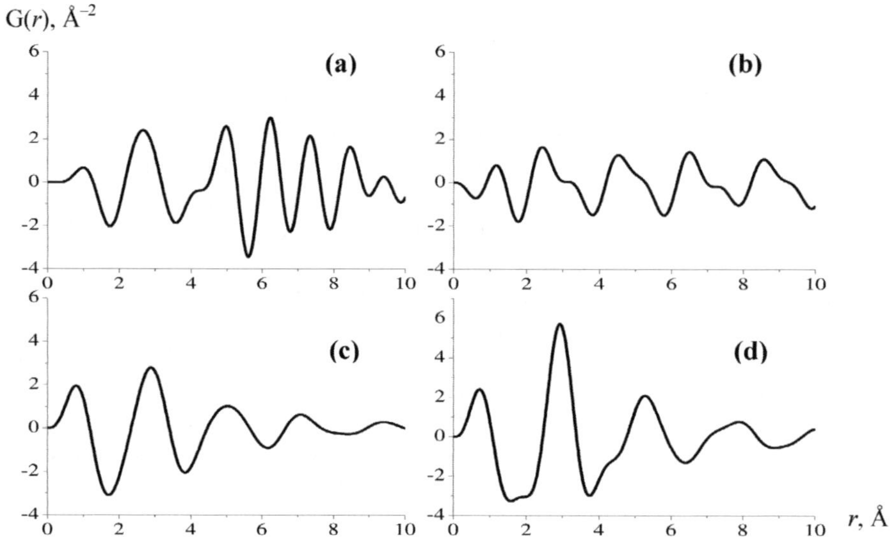

Figure 5. Radial distribution functions *G*(*r*) for DLC coatings alloyed with VIA group metals: (a) – *a*-C:H:Cr (100 % C_2H_2); (b) – *a*-C:H:Cr (20 % C_2H_2 + 80 % N_2); (c) – *a*-C:H:Mo:Si; (d) – *a*-C:H:W.

4. Conclusions

The results of this investigation have confirmed the correlation of the structural state of diamond-like carbon coatings alloyed with VIA group transition metals with the technology of their deposition as well as with their tribological properties.

Nanocomposite DLC coatings alloyed with chromium whose structure consists of metallic inclusions > 10 nm in size and nanometer chromium carbide and nitride inclusions have demonstrated the highest tribological characteristics. The DLC coatings alloyed with Mo and W had an amorphous type structure and their friction and durability parameters were considerably lower as compared with Cr-DLC.

Acknowledgements

The authors are grateful to M V Atamanov and G G Kirpilenko for coating deposition, and appreciate the financial support from the "MISiS Increase Competitiveness Program" of the Ministry of Education and Science of the Russian Federation (in part of nanohardness measurements) and from the Russian Foundation for Basic Research, project no. 15-08-05264-a.

References

[1] Levin I S, Khrushchov M M, Marchenko E A and Avdyukhina V M 2016 *Moscow Univ. Phys. Bull.* **71** 186–92

[2] Khrushchov M M, Atamanov M V, Marchenko E A, Petrzhik M I and Levin I S 2014 *Bull. Russ. Acad. Sci. Phys.* **78** 1007–14

[3] Khrushchov M M and Sveshnikov S V 2012 *Nanoinzheneriya* **8** 37–40

[4] Andersson M, Högström J, Urbonaite S, Furlan A, Nyholm L and Jansson U 2012 *Vacuum* **86** 1408–16

[5] Juhás P, Davis T, Farrow C L and Billinge S J L 2013 *J. Appl. Cryst.* **46** 560–6

Phase composition and residual stresses in thermal barrier coatings

A A Lozovan[1], S Ya Betsofen[1], A A Ashmarin[2], B V Ryabenko[3] and S V Ivanova[4]

[1] Moscow Aviation Institute (National Research University), ul. Orshanskaya 3, Moscow, 121552, Russia
[2] Institute of Metallurgy and Materials n. a. A A Baykova, Russian Academy of Sciences, Leninsky pr. 49, Moscow, 119991, Russia
[3] FGUP NPTsG Salyut, Moscow, 105118, Russia
[4] National Research Nuclear University MEPhi (Moscow Engineering Physics Institute), Moscow, 115409, Russia

E-mail: loz-plasma@yandex.ru

Abstract. X-ray study of the phase composition and residual stresses distribution in two-layer APS coatings showed that the ceramic layer consists of *t*-ZrO_2 phase with tetragonal lattice and the metal underlayer γ–solid solution based on nickel. In the transition zone thickness of ~ 100 µm as the distance from the surface was revealed a gradual transition from *t*-ZrO_2 to γ–solid solution. Increase in the specific volume of the metal underlayer resulting TGO growing leads to the formation of this layer high compressive stresses up to 600 MPa. In this case, the ceramic layer contains tensile stress up to 200 MPa.

1. Introduction

The studies performed in the last two decades in this field were directed toward improvement of the structure and properties of coatings with a ZrO_2 based external ceramic layer, which decrease the substrate material temperature by 100–150 °C [1, 2]. Alloying of ceramic with 7–8 wt. % Y_2O_3 stabilized the tetragonal modification of ZrO_2 and provided the formation of the *t'* phase, which is stable within several hundred thermal cycles in the temperature range up to 1200 °C.The resource of TBC depends substantially on the residual stresses, which are very high at the boundary between a ceramic layer and the metallic sublayer. It mainly consists of thermal growing oxide (TGO) Al_2O_3, which is the product of oxidation of a metallic MCrAlY sublayer. The purpose of this work is to study the distribution of phase composition, structure, and residual stresses in TBC by X-ray diffraction, scanning electron microscopy, and electron-probe microanalysis of the chemical composition in coating layers.

2. Experimental

Multilayer TBCs were deposited onto an ZhS26 alloy substrate in the following sequence: a metallic sublayer, a gradient (metal ceramic) intermediate layer, and a basic (external) ceramic layer. The external ceramic layer, the intermediate gradient layer, and the part of the metallic sublayer adjacent to it were deposited by plasma spraying in air in an automatic UniCoat (Sulzer Metco) device. The phase composition and the residual stresses in TBC were studied by X-ray diffraction (XRD) using a vertical

XRD-6000 SIMADZU device and CuK_α radiation (wavelength λ = 0.154178 nm). Crystalline phases were identified using the ICDD-2003 database.

3. Results and discussion

The X-ray diffraction pattern of the surface layer in the coating consists of lines of the tetragonal phase and only traces of the monoclinic phase (figure 1). This phase composition is retained to a depth of 180 μm. The X-ray diffraction patterns of the coating layers at a depth of 200–270 μm from the surface contain the lines of two phases, namely, t-ZrO_2 with yttria and an FCC nickel-based γ–solid solution (figure 2). This part of the coating corresponds to the intermediate gradient layer. The t-ZrO_2 phase is predominant in the part of the layer adjacent to the ceramic layer (figures 2(a), 2(b)), and a nickel-based γ–solid solution with an FCC lattice parameter a = 0.358–0.359 nm is predominant in the part next to the metallic sublayer (figure 2(d)). Note the presence of a satellite line near the (111) line of the γ–solid solution in all X-ray diffraction patterns of the intermediate layer. It can be attributed to an analogous line from pure nickel with a = 0.352 nm or to the (110) reflection of the β–phase (NiAl with the B2 lattice).

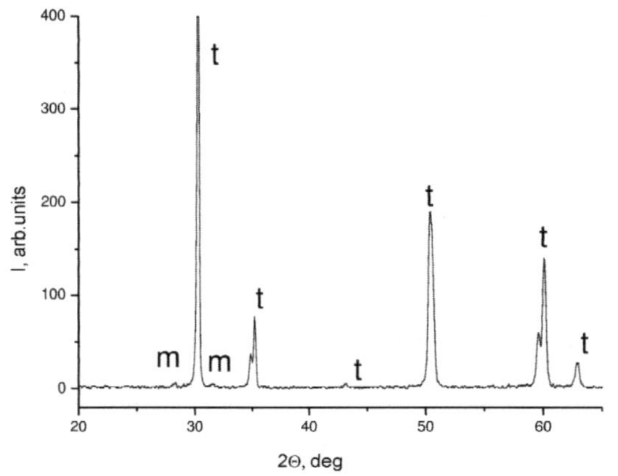

Figure 1. X-ray diffraction pattern of the surface of the ceramic layer.

The presence of the β–phase in this layer is possible, since the γ–solid solution in the intermediate layer adjacent to the external ceramic layer has the higher aluminum content as compared to the metallic sublayer. However, the (200) reflection of this phase (which has the highest intensity) is absent in all X-ray diffraction patterns of the metal ceramic layer that contain the satellite line, and the position of this line could correspond to the (110) reflection of the β–phase (see figures 2(b), 2(c)). Since the line intensities of the γ–solid solution indicate the absence of a pronounced texture, it is unlikely that the β–phase has such a strong texture that even traces of the most intense reflection are absent in an X-ray diffraction pattern. Thus, this satellite line should be attributed to a γ–solid solution depleted of alloying elements with a lattice parameter close to that of pure nickel. The presence of this line indicates that the oxidation of the γ–phase with the formation of TGO (Al_2O_3) begins at the stage of coating deposition. The oxidation of theγphase proceeds selectively, by the extraction of aluminum from the γ–solid solution. This is accompanied by the aluminum depletion of the γ–solid solution, at decrease in the lattice parameter from 0.359 to 0.352 nm, and the formation of Al_2O_3 in the intermediate layer next to ceramic. This behavior is reflected in X-ray diffraction patterns in the form of the satellite line near the main reflection of the γ–solid solution at an angle $2\theta \approx 44°$, which is shifted toward high angles ($2\theta \approx 45°$) from this reflection (see figures 2(b)–2(d)). The results of layer-by-layer analysis are generalized as the TBC (except for the ceramic layer) depth profiles of the lattice parameter of the nickel solid solution (figure 3). These results support the idea of the structural scheme of TBC, where a gradient metal ceramic layer forms between the metallic sublayer and the external ceramic layer. This gradient layer forms upon spraying of a ZrO_2–$8Y_2O_3$ ceramic powder preliminarily clad by nickel and aluminized. The intermediate layer is characterized by higher lattice parameters as compared to the metallic sublayer. This finding points to a higher aluminum concentration in the intermediate layer, which should increase the heat resistance of TBC and its service life. The most important role of the gradient layer is likely to be the fact that metal oxidation with the formation of TGO occurs in this layer, where metallic component particles are located discretely rather than forming extended layers. In the absence of the gradient layer, TGO forms at the interface between the ceramic and the metallic sublayer in the form of a continuous interlayer, which leads to the separation of ceramic when TGO reaches the critical thickness. Therefore, in the presence of the gradient intermediate layer, TBC fails significantly later and the service life of the structure increases.

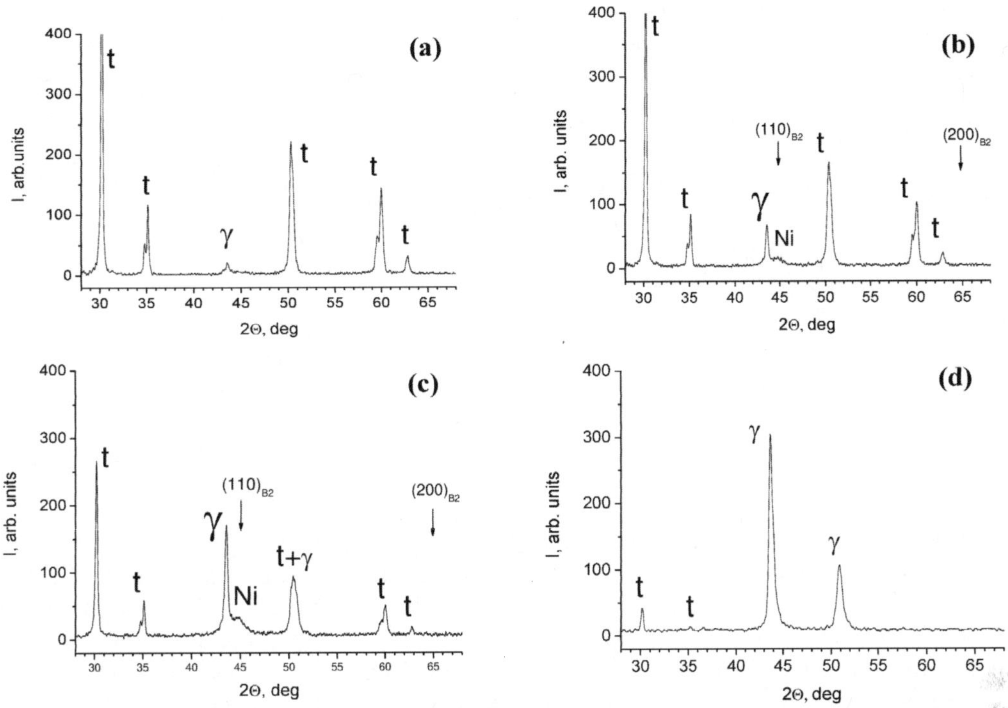

Figure 2. X-ray diffraction patterns of the layers in TBC that are located between the ceramic layer and the metallic sublayer at a distance from the coating surface, mm: (a) – 0.2; (b) – 0.22; (c) – 0.24; (d) – 0.27.

The process of coating deposition implies a nonuniform composition across the coating thickness and, hence, a lattice parameter gradient. Therefore, the $\sin^2\psi$ method becomes ineffective to determine the residual stresses. Figure 3 shows how the lattice parameter of solid solution changes with depth. The conditions of recording X-ray diffraction patterns in the $\sin^2\psi$ method are such that the lattice parameter gradient introduces substantial errors in the determination of the residual stresses. Therefore, to determine the residual stresses, we propose to use the technique [3], which allows us to avoid the errors related to the variation of the lattice parameter across the surface layer thickness. This technique is based on measuring the interplanar spacings for reflections ($h_1k_1l_1$) and ($h_2k_2l_2$), which exhibit the strongest difference between the elastic constants.

For cubic and tetragonal crystals, the residual stress is calculated from the relationship:

$$\sigma_{res} = \frac{a_{h_1k_1l_1} - a_{h_2k_2l_2}}{2(a_{h_2k_2l_2}K_{h_1k_1l_1} - a_{h_1k_1l_1}K_{h_2k_2l_2})}, \tag{1}$$

where $a_{h_1k_1l_1}$ and $a_{h_2k_2l_2}$ are the lattice parameters determined from the interplanar spacings for reflections ($h_1k_1l_1$) and ($h_2k_2l_2$).

Elastic constants $K_{hkl} = (\nu / E)_{hkl}$ are calculated using equations (2) and (3), respectively, for the cubic and tetragonal crystals:

$$\left(\frac{\nu}{E}\right)_{hkl} = -s_{11} - (s_{11} - s_{12} - 0.5s_{44})(h^2k^2 + h^2l^2 + k^2l^2)/(h^2 + k^2 + l^2), \quad (2)$$

$$\left(\frac{\nu}{E}\right)_{hkl} = s_{11}(h^2l^2 + k^2l^2 + 2h^2k^2) + (s_{33} - s_{44})(h^2l^2 + k^2l^2) + s_{12}(h^4 + k^4 + h^2l^2 + k^2l^2). \quad (3)$$

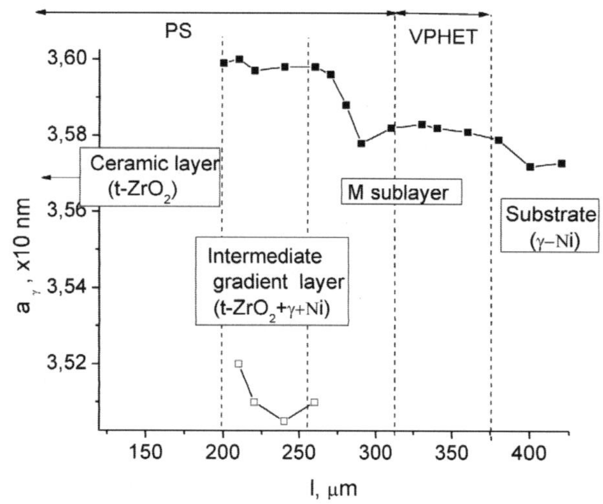

Figure 3. Depth profile of the lattice parameter of the γ–solid solution across TBC. PS is plasma spraying, and VPHET is vacuum plasma high-energy technology.

It is well known that sequential removal of surface layers occur relaxation of residual stresses which must be considered [4]. If the surface layer contains residual stresses, e. g. tensile stresses, they affect the remaining cross section of the sample via their axial and bending components, which are compensated by the axial stresses and moments of the opposite sign in the remaining cross section of the sample. Therefore, layer removal changes the stress distribution in the remaining part of the sample. When a layer of thickness a is etched off, the relaxed stresses are found by the summation of the stresses for all of the layers removed earlier. The stress in the removed layer ($\Delta\sigma_a$) of thickness a can be obtained by summation over all sequentially removed layers Δx_i:

$$\Delta\sigma_a = \sum \left[\Delta x_i \frac{\sigma_{i-1} + \sigma_i}{2} \left(\frac{1}{h - x_i} + 3\frac{h - x_i + \Delta x_i}{(h-a)^2} \right) \right].$$

The stress in a layer located at a distance a from the surface is then determined by the relation:

$$\sigma_a = \sigma^{meas} - \Delta\sigma_a.$$

Figure 4 shows the distribution of the experimental (σ^{meas}) and adjusted (σ_a) values of residual stresses over the entire cross section of TBC. According to these data, compressive stresses (600 MPa) are only present in the metallic sublayer, which is likely to be related to the formation of Al_2O_3 (which occupies larger atomic volume as compared to the metal). Peak tensile stresses are shifted by 50–100 μm from the ceramic layer to the gradient layer, i. e. from brittle ceramic to a more ductile metal ceramic. Stresses are absent in a significant part of the ceramic layer (figure 4).

4. Conclusions

An X-ray method of determining residual stresses based on the features of elastic anisotropy of the elastic characteristics of the ceramic and metal components of the TBC.

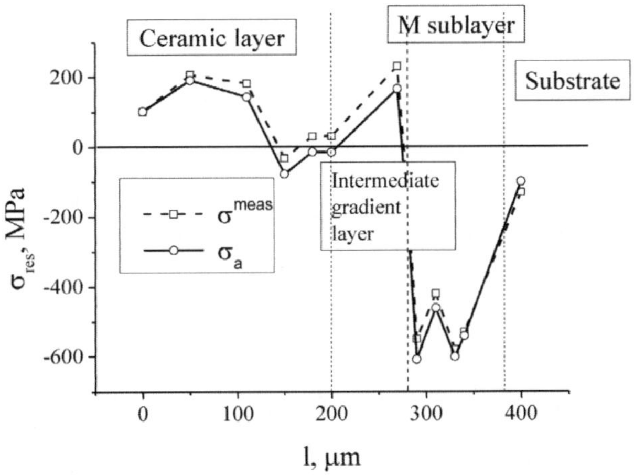

Figure 4. Distribution of residual stresses across TBC. PS is plasma spraying, and VPHET is vacuum plasma high-energy technology.

We proposed a technique for the correction of the residual stresses measured by an X-ray method along the depth of a coating with allowance for their relaxation upon the removal of surface layers.

The effectiveness of techniques demonstrated by TBC, consisting of a ceramic outer layer ZrO_2–$8Y_2O_3$ and the metal underlayer of an alloy NiCrAlY.

It is shown that the outer ceramic layer formed tensile stresses reach values of 200 MPa, which are likely the origin of quench, thus forming a metallic underlayer compressive stresses which reach 600 MPa, which are associated with an increase in specific volume due to the formation TGO.

Acknowledgments
The study was performed by a grant from the Russian Science Foundation (project №15-19-00241).

References

[1] Cernuschi F, Bianchi P, Leoni M and Scardi P 1999 *J. Therm. Spray Technol.* **1** 102
[2] DeMasi-Marcin J T and Gupta D K 1994 *Surf. Coat. Technol.* **68–9** 1
[3] Betsofen S Ya, Bannykh I O and Sarychev S M 2006 *Russian Metallurgy (Metally)* **5** 385
[4] Betsofen S Ya, Plikhunov V V and Ashmarin A A 2008 *Russian Metallurgy (Metally)* **2** 148

Journal of Physics: Conference Series **729** (2016) 012018 doi:10.1088/1742-6596/729/1/012018

Subminiature eddy current transducers for studying boride coatings

S F Dmitriev, A V Ishkov, V N Malikov and A M Sagalakov

Altai State University, Barnaul, Russia

E-mail: osys@me.com

Abstract. Strengthening of parts and units of machines, increased reliability and longer service life is an important task of modern mechanical engineering. The main objects of study in the work were selected steel 65G and 50HGA, wear-resistant boride coatings ternary system Fe–B–Fe_nB which were investigated by scanning electron microscopy and eddy-current non-destructive methods.

1. Introduction

In modern industry, enormous attention is paid to improving the reliability of parts and increasing their service life. Strengthening of parts and units of machines, increased reliability and longer service life is an important task of modern mechanical engineering. There are two different approaches to resolve this issue: the use of high-strength and wear-resistant materials and the processing of existing traditional materials can significantly increase the strength and other characteristics of the components. When processing materials with a view to improving operational characteristics of components, occupies an important place receiving and research of wear-resistant coatings suitable for use in the movable conjugations and working elements machines.

Currently, greatest distribution was received the hardening of details by means of induction hardfacing hard alloys and white cast iron. Nevertheless, the results of tests performed in the work – show that the most promising class of materials with a maximum erosion resistance is the hard coating (H > 10 GPa) made on the basis of metal carbides. However, among the factors influencing the erosion resistance, relates the thickness of the applied coating, its modes of deposition and the quality of the preliminary surface preparation. For example, for a titanium alloy, decrease the coating thickness to less than 15 μm led to a significant reduction in the erosion resistance.

As an promising materials for protective-strengthening coatings, offered matrix composites based on the ternary system Fe–B–Fe_nB, formed directly on the surface of the reinforcing parts, when it boriding in conditions of induction heating of the charge of the original composition and the various functional fillers.

Coating the ternary system Fe–B–Fe_nB formed by induction heating on the surface of the reinforcing parts when passing exothermic topochemical reaction between iron steel and boron the charge, the reaction captures the surface layer of the base-material and the reaction products form with it a single whole and characterized by smoothly changing the chemical composition of the transition at the interface base-coat that determines their high adhesion strength and wear resistance, special properties.

2. Materials used and the methods

The main objects of study in the work were selected steel 65G and 50HGA, wear-resistant boride coatings ternary system Fe–B–Fe_nB which were investigated by scanning electron microscopy (Philips SEM-515) and eddy-current non-destructive methods (ECT) [1, 2].

3. Materials used and the methods

The studied compositions of two-component mixture for boriding, applied to the steel specimens 65G, 50HGA as composition based on a liquid glass, the hydrolyzed ethylsilicate or polymer-based for boriding using induction heating, are presented in table 1.

Table 1. Formulations composition for wear resistant coatings using induction heating.

Composition	Component 1/mass %	Component 2/mass %/base component
1	B_4C/92	P–0,66/8
2	B_4C/84	P–0,66/16

Four main types of structures wear-resistant boride coatings produced using induction heating, have been found (figure 1): (a) – in the form of iron-boride eutectic carbide with closed regions; (b) – coating consisting of boride manganese crystals in the form of plates arranged in a softer matrix of a ledeburite-like, iron-boride eutectic; (c) – structure in the form iron-boride eutectic with large grains; (d) – a covering consisting of needle crystals manganese boride or chromium, disposed in a softer matrix of iron-boride ledeburite-like eutectic.

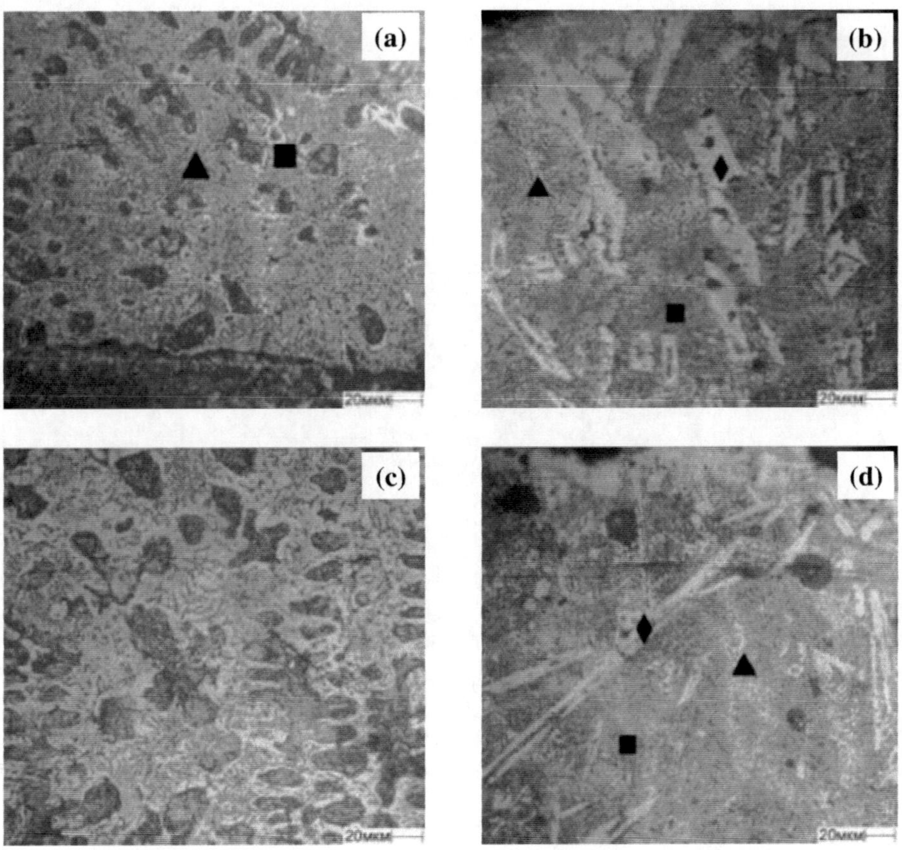

Figure 1. Types of structures wear-resistant boride coatings.

On figure 2(a) is represented by microhardness distribution in thickness boride coating obtained by using the composition 2 for different times, which shows that the coating is obtained for 76 and 106 s, has a higher microhardness (HV100 13 770 MPa) as compared with coatings obtained in 120 and 136 s, or, for example, a coating formed from the composition 1 at the same time boriding. Figure 2(b) shows the results of wear resistance tests boride coatings obtained on steel 65G using induction heating in the formulations 1 and 2 for different times.

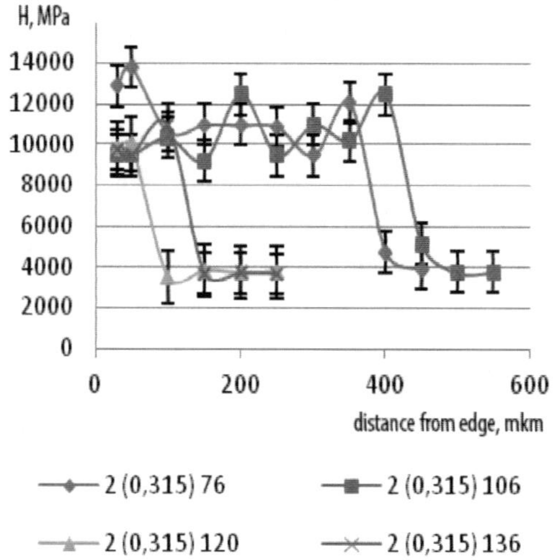

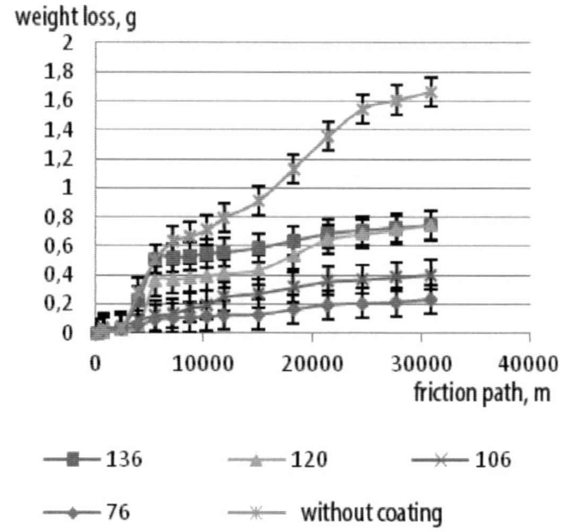

Figure 2. Microhardness distribution in thickness (a) and the wear resistance of boride coatings on steel 65G (b) obtained from formulations 1, 2 (table 1) for different times.

Scanning the sample was carried out using two differentially connected sensors. Scan scheme is shown in figure 3.

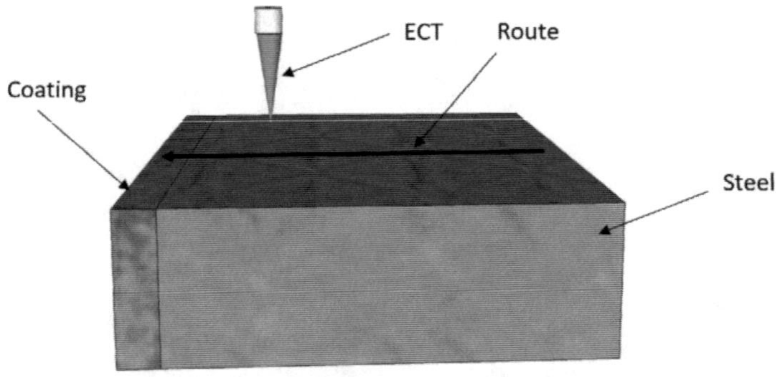

Figure 3. Scan scheme.

Figure 4 shows the results of studies of coatings produced using the eddy-current method. A clear link electromagnetic and strength characteristics of the coating was established. When scanning the sample with no coating, the rate of fall of the signal amplitude, carrying information about the properties of the substance considerably exceeded the rate of fall of the signal when the sample is scanned with coatings. Comparison of test results shows the presence of the dependencies between the electromagnetic characteristics and wear resistance of coatings.

Increased wear resistance of hardened layers with increasing of time induction heating in the composition 2 (table 1) explained by a decrease of hardness the main strengthening phase. This permits asserting that the characteristics of boride coatings obtained on the structural and alloyed steels using induction heating, do not change monotonically and have a certain optimum in their function of time.

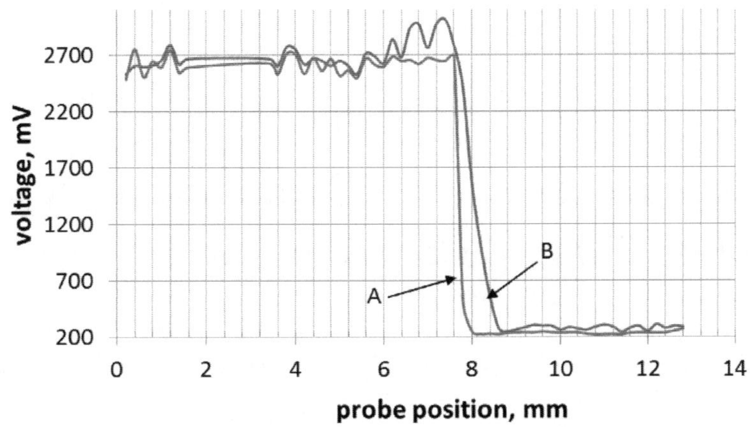

Figure 4. Dependences eddy-current sensor signal amplitude.

To control these characteristics, is advisable to use non-destructive eddy-current methods to promptly investigate the resulting coatings and draw conclusions about their quality.

References

[1] Malikov V N, Dmitriev S F, Sagalakov A M and Ishkov A V 2014 *Instruments and experimental techniques* **6** 751–4

[2] Dmitriev S F., Katasonov A O, Malikov V N and Sagalakov A M 2016 *Russian Journal of Nondestructive Testing* **1** 32–7

Journal of Physics: Conference Series **729** (2016) 012019 doi:10.1088/1742-6596/729/1/012019

Method of magnetron target temperature evaluation by analysis of thermal radiation spectrum

V A Gerasimov, D K Kostrin, L M Selivanov, A A Uhov, A E Komlev and E S Shutova

St. Petersburg Electrotechnical University "LETI", 197376, St. Petersburg, Russia

E-mail: aauhov@yandex.ru

Abstract. A method of vacuum chamber magnetron target temperature evaluation by measurement of aggregated spectrum of gas discharge emission and target thermal radiation is proposed. Method is based on matching of calculated black body radiation spectrum with given temperature and the minimal points on the measured spectrum.

In the few last years has greatly increased interest in the study of high-power metal targets magnetron sputtering process [1–3]. One of the main factors of high-power sputtering process is the considerable heating of target up to very high temperatures. It is very important to have a reliable instrument for target temperature measurement right during the sputtering process. Direct measurement by means of contact sensor is difficult because of fast temperature destruction of such sensors. Traditional contactless pyrometric methods give a significant measurement error due to the hot target sputtering process output optical radiation is the result of two different processes – a gas discharge light emission and the target thermal radiation. Object temperature measurement by means of optical spectrum processing (spectrum pyrometry) now is well known [4, 5] and became more popular due to cost reduction and availability of compact spectrometer devices. A main trouble of using spectrum pyrometry for hot magnetron target temperature measurements is the big quantity of intensive emission spectral lines in the input spectrum that makes the task of thermal radiation spectrum segregation a rather complex (figure 1). A given article describes a method of spectrum measurements processing for target temperature evaluation.

The experiments were performed at the modernized setup UVN-71 equipped with 78 dm^3 vacuum chamber and flat magnetron. A target was made of round tantalum sheet with a diameter of 70 mm and thickness of 0.5 mm. To provide a hot target mode, the tantalum target was fixed with gap of 1 mm to the metal plate, cooled by flowing water. A target sputtering was performed with the residual pressure of 10^{-2} mTorr, argon pressure of 3.5 mTorr and discharge current density from 8 to 36 mA/cm^2.

For the spectrum measurement a compact spectrometer device ISM3600 was used [6, 7]. ISM3600 is capable for optical spectrum measurement in region 250–1000 nm with the resolution in visible area not worse than 2.5 nm and absolute wavelength measurement error not greater than ±0.5 nm. The optical signal output from vacuum chamber is provided by quartz monofiber with diameter of 0.4 mm. ISM3600 software has a database with periodic table elements spectral lines that allows detecting the elemental and ionic composition of gas discharge during the magnetron sputtering.

Journal of Physics: Conference Series **729** (2016) 012019 doi:10.1088/1742-6596/729/1/012019

Temperature calculation method is based on separation of target thermal radiation spectrum from the total light spectrum gathered from the chamber. The separation is performed in few stages:

• depending on the number of emission spectral lines, a viewing window be defined; the smaller the lines number, the wider the viewing window and vice versa. For most of the plasma emission spectra, the window size is approximately of 30 nm;

• inside the first viewing window, at begin of spectral data, a minimum value (minimum signal level) be sought and stored it in the buffer. Then algorithm switches to the next viewing window and repeats the operation along the all spectral data, received from the spectrometer device;

• using the array of minimums obtained at the previous step, a thermal radiation spectrum being constructed by third-degree polynomial approximation.

The source unfiltered data and the resulting spectrum obtained for experiment with discharge power density 14.8 W/cm^2 are shown on figure 2.

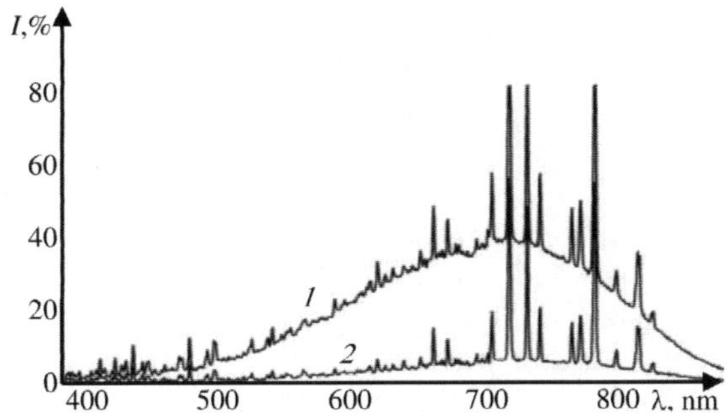

Figure 1. Optical spectrum obtained from vacuum chamber with magnetron and hot tantalum target with discharge power density: *1* – 14.8 W/cm^2; *2* – 11.3 W/cm^2.

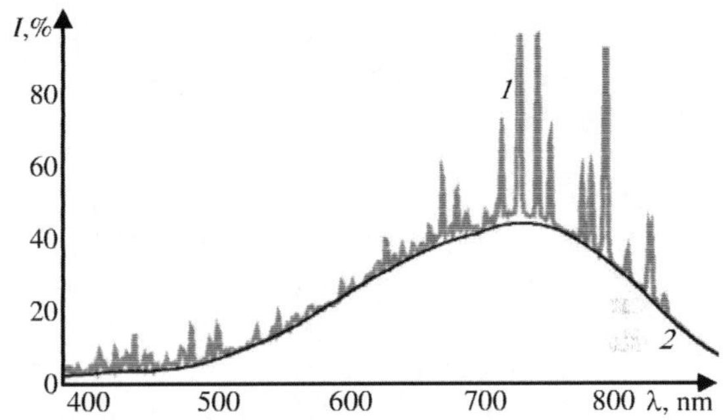

Figure 2. Thermal radiation spectrum separation: *1* – input spectral data; *2* – separated thermal radiation spectrum.

For target's temperature evaluation a spectrum of black body radiation with average operational temperature in the given spectrum range is being calculated. A calculation of spectral intensity is processed by well-known Planck's formula:

$$I(\lambda) = \frac{2\pi hc^2}{\lambda^5\left(e^{hc/\lambda kT} - 1\right)},$$

where $I(\lambda)$ – spectral intensity at the given wavelength λ; h – Plank's constant; k – Bolzman's constant; T – object's temperature (K); c – light speed in vacuum.

The calculated intensity $I(\lambda)$ is normalized to the value of the measured thermal radiation intensity at a fixed wavelength of 850 nm, and then the integral difference between the two spectra – theoretically for a selected temperature, and segregated from the real spectrum is computed. Such difference is calculated only for the signal ranges where intensity is not less than preset threshold (for example, 5 % from the maximum). Then calculations are repeated for other temperatures and after few iterations (using, for example, a bisection method), it is possible to find a temperature value where the integral difference is minimal.

Example is shown on figure 3: a spectrum data are obtained from target, heated to 1680 K with discharge power density 14.8 W/cm^2; curve *1* corresponds to black body radiation spectrum with T = 1680 K. A deviation from this temperature to, for example, ±20 K gives significantly bigger integral difference (curves *2* and *3* on figure 3).

A set of experiments with different discharge power density were performed and target's temperatures were evaluated. The results show a linear dependence between power and temperature (figure 4) that corresponds to target heating process.

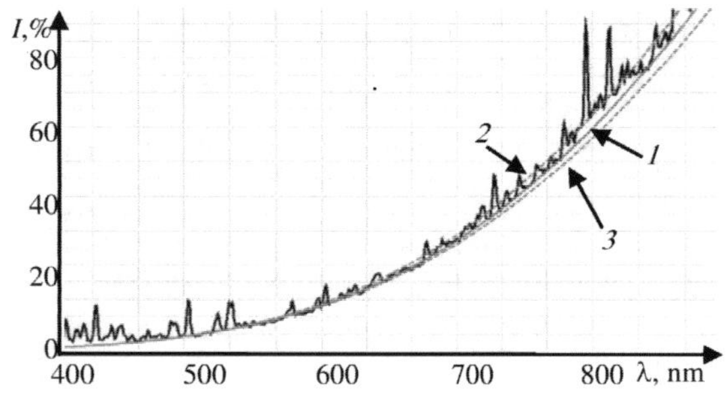

Figure 3. Calculation of black body radiation spectrum.

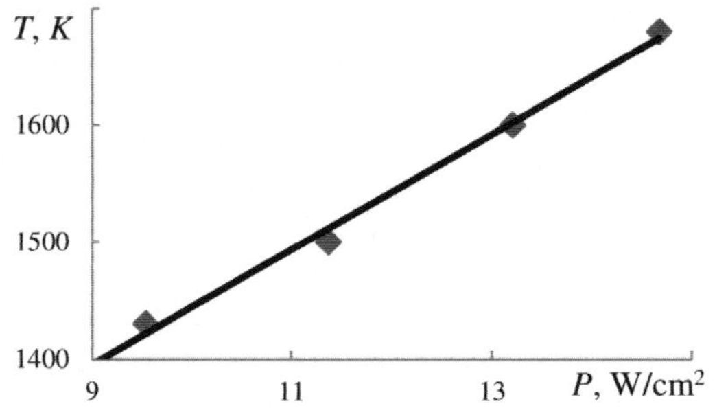

Figure 4. Hot target temperature vs. discharge power density.

One of given method obvious disadvantage is impossibility of thermal radiation accurate segregation due to errors in signal minimums detection caused by the optical sensor noise. This can be compensated by applying a various algorithms of signal smoothing.

References

[1] Lapshin A E, Levitskii V S, Shapovalov V I *et al* 2016 *Glass Phys. Chem.* **42** (*in print*)
[2] Komlev A E, Shapovalov V I and Shutova N S 2012 *Technical Physics* **7** 1030–3
[3] Shapovalov V I, Komlev A E, Bondarenko A S, Baykov P B and Karzin V V 2016 *Physics Letters, Section A: General, Atomic and Solid State Physics* **7–8** 882–5
[4] Zhang Z M, Tsai B K and Machin G 2010 *Radiometric temperature measurements. I. Fundamentals. Experimental methods in the physical sciences. V. 42* (Amsterdam: Elsevier)
[5] Magunov A N 2009 *Instruments and Experimental Techniques* **4** 451–72
[6] Uhov A A, Gerasimov V A, Kostrin D K and Selivanov L M 2014 *Journal of Physics: Conference Series* **567** 012039
[7] Kostrin D K, Uhov A A and Lisenkov A A 2016 *Vakuum in Forschung und Praxis* **3** 34–7

Study of the phase composition of silicide coatings, based on layered Nb–Mo structures, obtained by vacuum-arc deposition

A A Lozovan[1], **S Ya Betsofen**[1] **and A S Lenkovets**[2]

[1] Moscow Aviation Institute (National Research University), ul. Orshanskaya 3, Moscow, 121552, Russia
[2] JSC «Komposit», Korolev, Moskovskaya obl., 141070, Russia

E-mail: loz-plasma@yandex.ru

Abstract. A multilayer composite ~1000 μm in thickness, formed by niobium and molybdenum layers (number of layers n = 230), is obtained by vacuum-arc deposition with subsequent siliconization of the surface layers at a temperature of 1200 °C. Layer-by-layer phase analysis is performed by X-ray diffraction and scanning electron microscopy. It is found that in the surface layers ~130 μm in thickness, single-phase silicides $(Nb_xMo_{1-x})Si_2$ are formed with the hexagonal C40 structure (Strukturbericht designations). Alternating layers of solid solutions based on niobium and molybdenum with a body-centered cubic (BCC) lattice are observed within the composite. The formation of solid solutions caused by heating of the coating leads to convergence of the values of the linear thermal expansion coefficient and Young's modulus at the interface between the layers.

1. Introduction

Multilayer coatings of alternating metal or ceramic layers are attracting increasing attention because of their excellent mechanical properties, such as high hardness, adhesion strength and toughness, and wear resistance, and relatively low levels of residual stress [1–3]. These properties are the result of deviation of the crack path due to the presence of a soft metal layer in conjunction with an obstacle in the form of a solidceramic layer; the ability to create favorable residual stress gradients; a decrease in the bending stress resulting from better load distribution in a layered material compared to a monolithic material; and the branching of cracks because of different values of the elastic moduli of the layers. Therefore, investigations of the formation of silicide coatings based on layered structures obtained by the ion-vacuum deposition of alternating layers of niobium and molybdenum, followed by saturation of the surface layers with silicon, are of considerable scientific and practical interest, which is selected as the topic of this paper.

2. Experimental

To obtain a multilayer coating, an NNV-6.6-I1 setup equipped with three vacuum-arc evaporators was selected. A 5VMTs niobium alloy and molybdenum (MCh grade) were used as evaporated materials (cathode materials). The coatings were deposited on a cylindrical substrate of copper (M1 grade) mounted in a mandrel. The chamber was evacuated to a residual pressure of $P = 0.33 \cdot 10^{-3}$ Pa; argon was introduced to a pressure of P = 3–4 Pa; and the surface of the samples was purified in the glow discharge plasma at a voltage of 1.1 kV for 30 min. After purification, the argon pressure was reduced $P = 4 \cdot 10^{-1}$ Pa, and the sample was heated by an electron beam. Then, layers of niobium and molybdenum were sputtered alternatively at a substrate bias voltage of $U = 40$ V to form a layered metal composite approximately 1 mm in thickness. The thickness of the molybdenum layer was 1 μm, and the niobium layer was 3–4 μm thick. The copper substrate was then removed by selective chemical etching, and the composite was subjected to diffusion siliconization at 1200 °C for 10 h.

3. Results and discussion

Figure 1 shows a cross-sectional structure of the multilayer coating, which testifies to the absence of cracks and bundles in alternating layers of molybdenum and niobium. Figure 1(b) shows a few traces of the droplet phase.

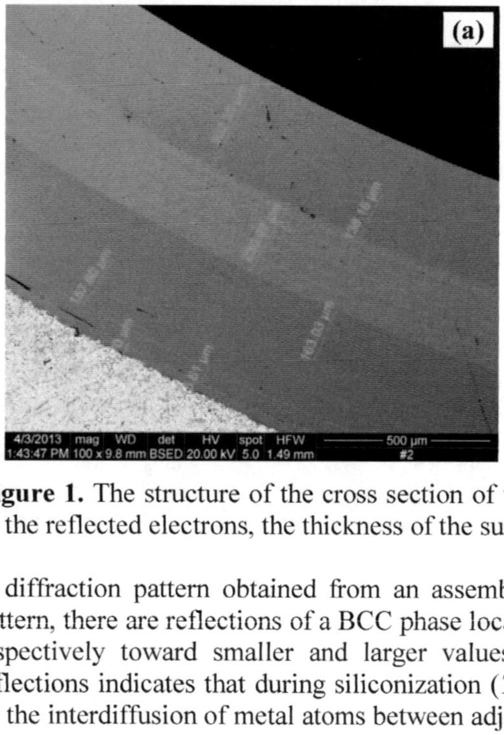

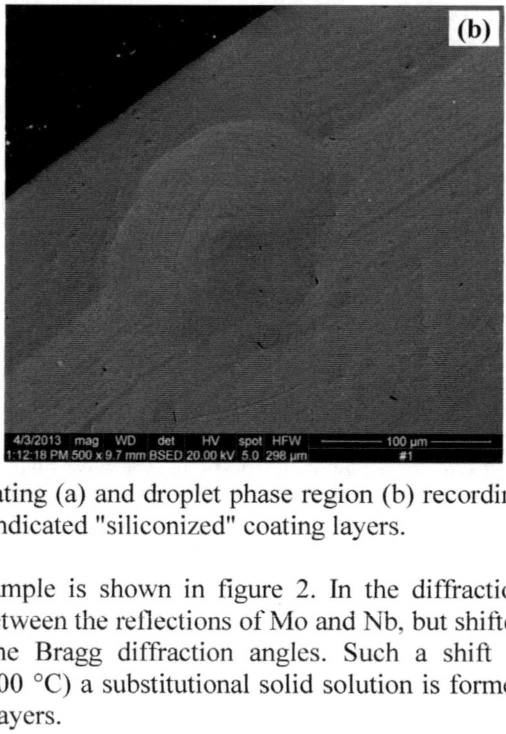

Figure 1. The structure of the cross section of the coating (a) and droplet phase region (b) recording in the reflected electrons, the thickness of the surface indicated "siliconized" coating layers.

A diffraction pattern obtained from an assembled sample is shown in figure 2. In the diffraction pattern, there are reflections of a BCC phase located between the reflections of Mo and Nb, but shifted respectively toward smaller and larger values of the Bragg diffraction angles. Such a shift in reflections indicates that during siliconization ($T = 1200$ °C) a substitutional solid solution is formed by the interdiffusion of metal atoms between adjacent layers.

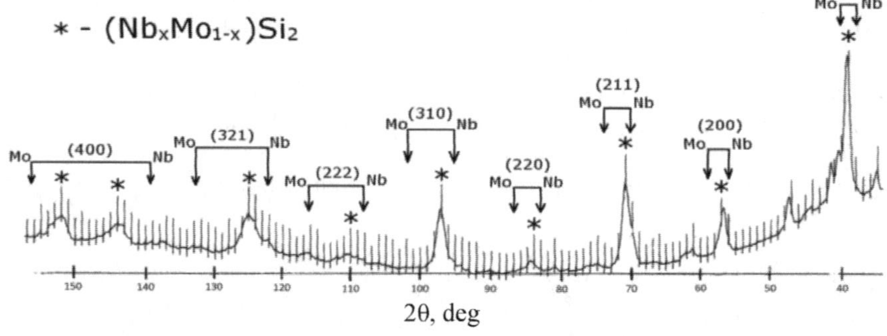

Figure 2. Diffraction pattern of the assembled sample of Nb–Mo solid solutions; arrows indicate the positions of the reflections of pure niobium and Mo.

In the diffraction pattern of the surface layer, there are all reflections of the $NbSi_2$ phase without exception corresponding to the ICDD-2003 database, and there are no extraneous reflections (figure 3), which is very rare in such complex non equilibrium systems as multilayer vacuum-arc coatings, particularly those subjected to high-temperature siliconization. Therefore, we can assume that there are no other silicides in the surface layer. The mean values of the composition of the solid-solution phases with a predominance of niobium (solid solution A) and molybdenum (solid solution B) were estimated by averaging over all available reflections in the diffraction pattern, taking into account the texture and proportions of the two phases for each reflection according to equation

$$C_{\mathrm{Mo}} = \frac{1}{n}\sum_{i=1}^{n}(C_{\mathrm{Mo}}^{\mathrm{A}} P_i^{\mathrm{A}} f_i^{\mathrm{A}} + C_{\mathrm{Mo}}^{\mathrm{B}} P_i^{\mathrm{B}} f_i^{\mathrm{B}}), \qquad (1)$$

where $C_{\mathrm{Mo}}^{\mathrm{A}}$ and $C_{\mathrm{Mo}}^{\mathrm{B}}$ are the molybdenum concentrations in solid solutions A and B, respectively; P_i^{A} and P_i^{B} are the pole densities of the *i*-th reflex in solid solutions A and B, respectively; f_i^{A} and f_i^{A} are the volume fractions of solid solutions A and B.

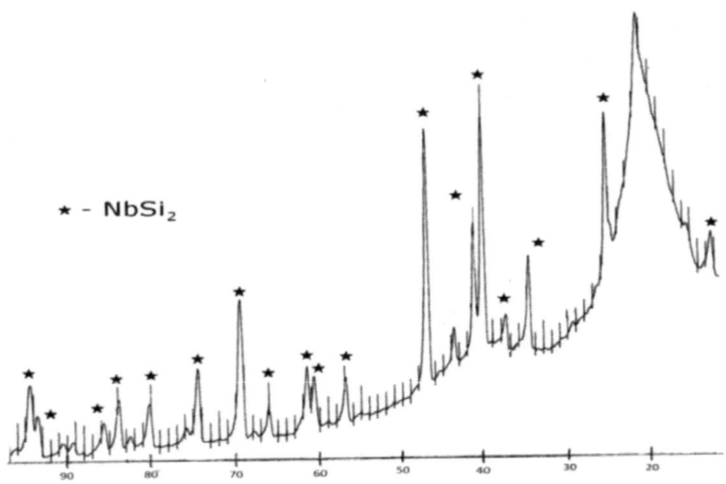

Figure 3. Diffraction pattern obtained from the outer surface of the sample after siliconization.

X-ray diffraction studies have shown that as a result of the diffusion siliconization at 1200 °C of a Nb–Mo layered composite (n = 230, T ~ 1000 μm) with alternating metal layers in the surface region ~130 μm in thickness, single-phase silicides $(Nb_xMo_{1-x})Si_2$ with the hexagonal C40 structure are formed. The single phase structure of the silicide, formed from bicomponent solid solutions of Mo and Nb, corresponds to the pseudobinary state diagram of $NbSi_2$–$MoSi_2$ and is characteristic of a number of similar systems $MoSi_2$–XSi_2 (where X = Ta, Ti, Ir, Zr, V or Cr) [4, 5]. All of these elements except molybdenum form silicides with the C40 structure, this structure is metastable only for $MoSi_2$, and the stable silicide forms the tetragonal C11b structure. The pseudobinary state diagrams of $MoSi_2$–XSi_2 are characterized by a wide homogeneity range of solid solutions of the C40 structure and relatively narrow two-phase and single-phase regions of the C11b structure. This explains the presence of the coating of single-phase silicide C40 in the surface layer, despite the significant concentration of molybdenum in the initial coating: the ratio of the metal components corresponds to the composition of $Nb_{0.62}Mo_{0.38}$.

4. Conclusions

A multilayer Nb–Mo coating (n = 230) with a thickness of ~1 mm is obtained by vacuum-arc deposition.

It was found that in the surface layers ~130 μm in thickness silicides $(Nb_xMo_{1-x})Si_2$ are formed with the hexagonal C40 structure.

The rest of the coating section by heating accompanying siliconizing process formed of alternating layers of BCC solid solutions based on Nb and Mo nominal composition $Nb_{0.75}Mo_{0.25}$ (78 %) and $Nb_{0.18}Mo_{0.82}$ (22 %), respectively.

Acknowledgments
The study was performed by a grant from the Russian Science Foundation (project №15-19-00241).

References

[1] Yanhui Z, Guoqiang L, Jinquan X, Hao D, Chuang D and Gao L 2011 *Applied Surface Science* **257** 2683
[2] Hovsepian P Eh, Ehiasarian A P, Braun R, Walker J and Du H 2010 *Surf. & Coat.Technol.* **204** 2702
[3] Pearson W B 1958 *A Handbook of lattice spacings and structures of metals and alloys* (NewYork: Pergamon Press)
[4] Hagihara K, Nakano T, Hata S, Zhu O and Umakoshi Y 2010 *Scripta Materialia* **62** 613
[5] Hagihara K, Hama Y, Nakano T, Yuge K and Nakano T 2013 *Acta Materialia* **61** 3432

Journal of Physics: Conference Series **729** (2016) 012021 doi:10.1088/1742-6596/729/1/012021

Probe measurements of magnetron discharge parameters

V Karzin and V Smirnov

St. Petersburg State Electrotechnical University "LETI", 5 Prof. Popov Street, St. Petersburg 194354, Russia

E-mail: karzin.ru@gmail.com

Abstract. Langmuir probe measurements have been carried out in a positive column discharge of a planar magnetron with a titanium target. In this work concentration and temperature of electrons is measured in positive column of magnetron discharge with Langmuir probe. Concentration of electrons is found to be on the level of 10^{16} m^{-3} at a notable distance (over 200 mm). Two groups of electrons exist: «cold» and «hot» with average temperature of 16 000 and 41 000 K respectively. Dependence of discharge parameters on working gas pressure is measured.

1. Introduction

Probing method of measuring of plasma parameters has been used for almost hundred years. It is applied in investigations of different parameters of gas discharges. The method is known to be a reliable instrument for measuring local temperature and concentration of charged particles and plasma electric potential [1–6]. In this work concentration and temperature of electrons is measured, in positive column of magnetron discharge, by a single probe.

2. Experimental details

Dependence of the parameters on gas pressure and current density is investigated. The experiments were performed using an oil-diffusion pumped vacuum chamber with a diameter of 550 mm and a height of 500 mm. A schematic of the experimental chamber is shown in figure 1.

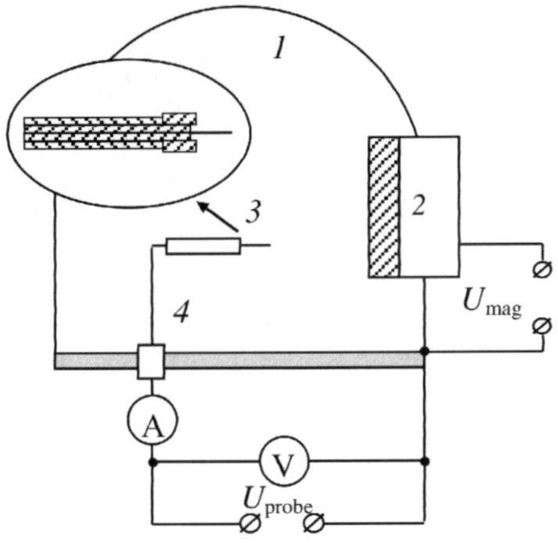

Figure 1. Schematic picture of discharge investigation by probing method: *1* – vacuum chamber (anode); *2* – magnetron; *3* – Langmuir probe; *4* – vacuum entrance.

Journal of Physics: Conference Series **729** (2016) 012021 doi:10.1088/1742-6596/729/1/012021

The chamber was evacuated prior to the measurements to a base pressure of 0.02 mTorr. Target is made of titanium (purity 99.99 %) and working gas is argon (purity 99.998 %). Argon gas was introduced through a leak valve so that the desired pressure of the sputtering gas was achieved. A 160 mm diameter Ti target was mounted on the magnetron cathode. The probe material is molybdenum wire 0.5 mm diameter and 4 mm length.

The probe (cylindrical electrode) was situated horizontally against the target's erosion area. Measurements were conducted at distances, where magnetic field's influence is negligible, and conditions are similar to plasma (positive column). This was confirmed by *I–V* measurements of the probe at distances of 200, 250 and 300 mm. Obtained results are different not more than by 10 %. That confirms homogenous distribution of positive column's parameters. To eliminate error, connected to polluting of collector with titanium, a removable head of the probe was periodically replaced.

3. Result and discussion

Figure 2(a) shows current-voltage curves of a single probe. Voltage was swept in the range of –50…+50 V. Current in the circuit is provided by negatively charged particles – electrons, if the probe potential is positive. Negative potential on the probe leads to ionic current, which is significantly less than electronic. Relatively high positive probe potential (over +40 V) leads to an intensive heating of the probe (its color changed to orange-red).

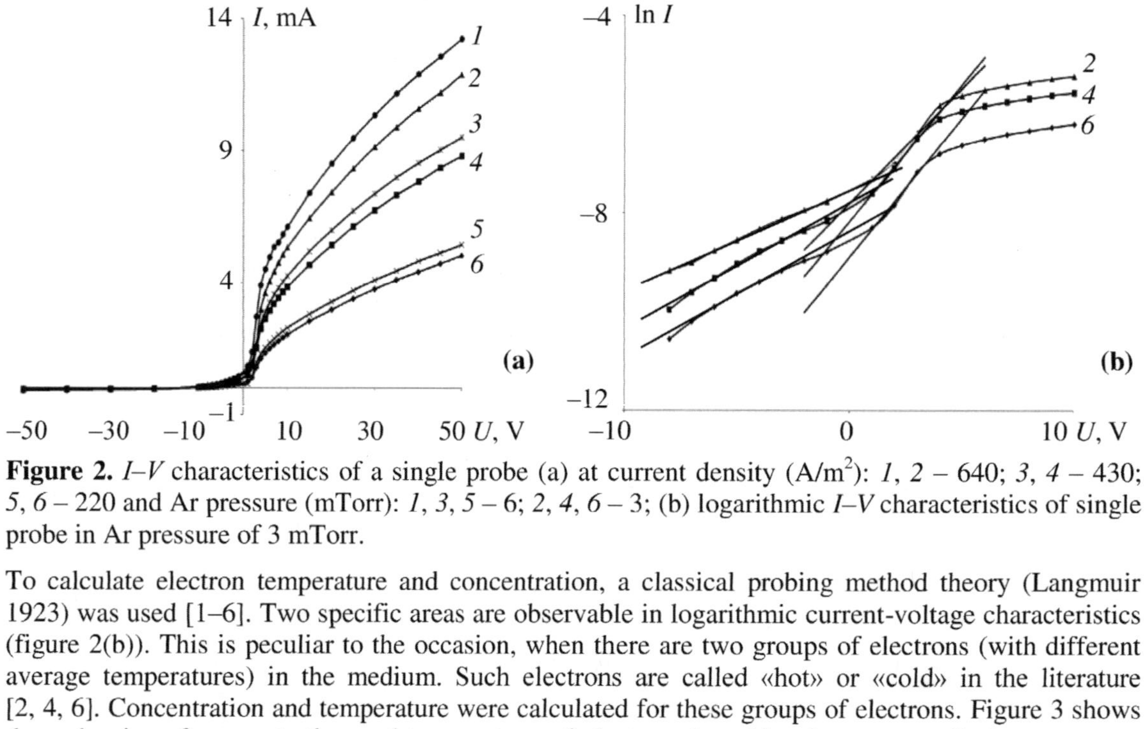

Figure 2. *I–V* characteristics of a single probe (a) at current density (A/m^2): *1, 2* – 640; *3, 4* – 430; *5, 6* – 220 and Ar pressure (mTorr): *1, 3, 5* – 6; *2, 4, 6* – 3; (b) logarithmic *I–V* characteristics of single probe in Ar pressure of 3 mTorr.

To calculate electron temperature and concentration, a classical probing method theory (Langmuir 1923) was used [1–6]. Two specific areas are observable in logarithmic current-voltage characteristics (figure 2(b)). This is peculiar to the occasion, when there are two groups of electrons (with different average temperatures) in the medium. Such electrons are called «hot» or «cold» in the literature [2, 4, 6]. Concentration and temperature were calculated for these groups of electrons. Figure 3 shows dependencies of concentration and temperature of «hot» and «cold» electrons on discharge current density for two different gas pressures.

Experimental data processing has shown two groups of electrons («hot» and «cold») in positive column of magnetron's discharge. Temperature of «cold» electrons (argon pressure 3 mTorr) raises from 16 500 to 20 200 K with current density increasing from 200 to 600 A/m^2 . Their concentration is increases from $8.8\cdot10^{15}$ to $2.2\cdot10^{16}$ m^{-3}. Temperature of «hot» electrons with same current density increasing (argon pressure 3 mTorr) raises from 42 000 K to 47 700 K and concentration raises from $5.5\cdot10^{15}$ to $1.4\cdot10^{16}$ m^{-3}.

As a result, two groups of electrons in positive column of magnetron's discharge were shown. Concentration of «cold» electrons is ~ 1.6 times higher in average than concentration of «hot» ones and their temperature is lower ~ 2.5 times. Amount of ionized gas atoms increases with the increase of discharge current density. That explains the increase of electron concentration inside the investigated medium. This can be clearly seen in presented figures. Increase of working gas pressure entails increase of electron concentration and decrease of their average temperature. This is applied to both «hot» and «cold» electrons.

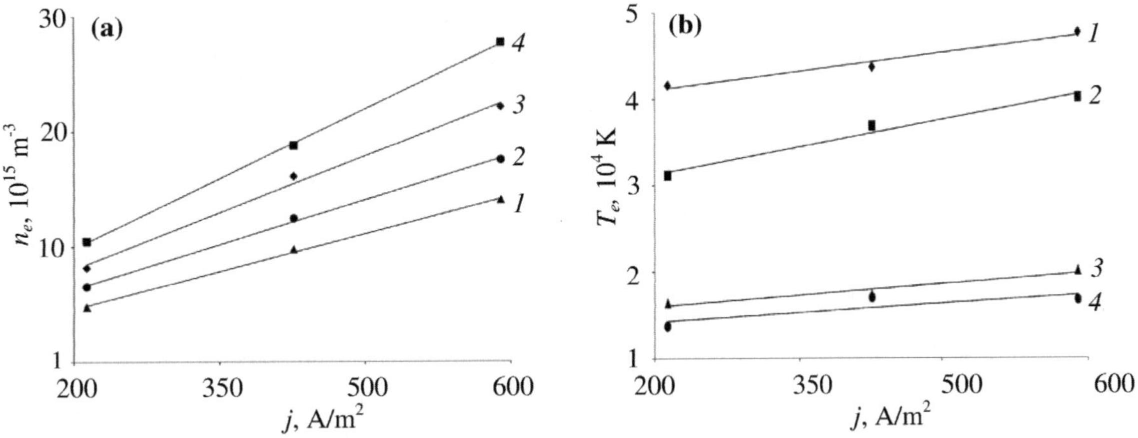

Figure 3. Dependence of concentration (a) and temperature (b) of «hot» (*1*, *2*) and «cold» (*3*, *4*) electrons on current density of magnetron discharge with different Ar pressure (3 mTorr (*1*, *3*) and 6 mTorr (*2*, *4*)).

4. Conclusions

In this preliminary study a Langmuir probe has been used to measure of the positive column parameters (T_e, n_e) in a DC magnetron discharge. Results show a number of interesting features, which warrant further investigation. Langmuir probe measurements have been performed in a planar magnetron system. The electron density, electron temperature, of «cold» and «hot» groups of electrons has been determined in various argon pressures in the range from 3 to 6 mTorr and current density in the range from 200 to 600 A/m^2.

The work makes it possible to determine an electron energy distribution function in the discharge. That can be used to develop a theoretical model of a gas discharge. It is possible to compare results, obtained with usual magnetron sputtering mode, to results, obtained with «hot target» sputtering mode and high-power pulsed sputtering.

Acknowledgments

The authors thank Russian Science Foundation for their support on the study (grant 15-19-00076).

References

[1] Langmuir I and Mott-Smith H M 1926 *Phys. Rev.* **28** 727–63
[2] Field D J, Dew S K and Burrell R E 2002 *J. Vac. Sci. Technol.* **20** 2032–41
[3] Spolaore M Antoni V and Bagatin M 1999 *Surface and Coatings Technology* **116** 1083–8
[4] Bradley J, Backer H, Kelly P and Arnell R 1999 *Surface and Coatings Technology* **135** 221–8
[5] Bystrov Yu A, Vetrov N Z and Lisenkov A A 2011 *Technical Physics Letters* **37** 707–9
[6] Bystrov Yu A, Vetrov N Z and Lisenkov A A 2010 *Technical Physics Letters* **36** 570–3

Journal of Physics: Conference Series **729** (2016) 012022 doi:10.1088/1742-6596/729/1/012022

Sensor for the working surface cleanliness definition in vacuum

E A Deulin[1], S S Mashurov[2] and A A Gatsenko[2]

[1] Moscow State Technical University, Russia
[2] CJSC "Aerospace monitoring and technologies" (CJSC "AMT"), Russia

E-mail: deulin@bmstu.ru

Abstract. Modern development of nanotechnology as one of the modern science priority directions is impossible to imagine without the use of vacuum systems and technologies. And the better the vacuum (lower the pressure), the "cleaner" we get a surface, which is very important for nanotechnology. Determination of the cleanliness of the surface or the amount of molecular layers of adsorbed gases on the working surface of the products especially in industry, where the cleanliness of the working surface is a key parameter of the technological process and has a significant influence on the output parameters of the final product is the main goal of this work.

Modern development of nanotechnology as one of the modern science priority directions is impossible to imagine without the use of vacuum systems and technologies. In vacuum we can deposit thin films, grow crystals for the semiconductor industry, manufacture of microelectromechanical systems (MEMS), investigation of the surface at the atomic level, etc. This is due to the fact that in vacuum we can obtain the atomically clean surface during the entire technological process, i. e. the surface where the amount of adsorbed gases will be negligible compared to atmospheric pressure.

And the better the vacuum (lower the pressure), the "cleaner" we get a surface, which is very important for nanotechnology. To prepare the product for further technological process in vacuum using traditional methods of working surface cleaning, such as the chemical treatment in atmosphere, the plasma dry cleaning, ion beam treatment, heat treatment in vacuum, etc. [1, 2]. However, such methods have their drawbacks and it is impossible to say with certainty that after such cleaning, the surface was "clean".

So how to determine the cleanliness of the surface or the amount of molecular layers of adsorbed gases on the working surface of the products especially in industry, where the cleanliness of the working surface is a key parameter of the technological process and has a significant influence on the output parameters of the final product? On the one hand is the use of adsorption isotherms. However, such calculations are time consuming and require the development of mathematical models for various modes of operation of the vacuum system [3].

On the other hand is to use the sensor that measures the cleanliness of the working surface of the product in vacuum, developed in Moscow State Technical University jointly with CJSC "AMT", analogues of which do not exist (figure 1). The main feature of the developed sensor is its versatility, i. e. the possibility of measuring the cleanliness of the working surfaces or amount of molecular layers of adsorbed gases on the surface of the product in the pressure range from 10^5 to 10^{-10} Pa (extremely high vacuum, XHV), in the temperature range from –40 to +150 °C and in the humidity range from 0.1 to 0.95 RH.

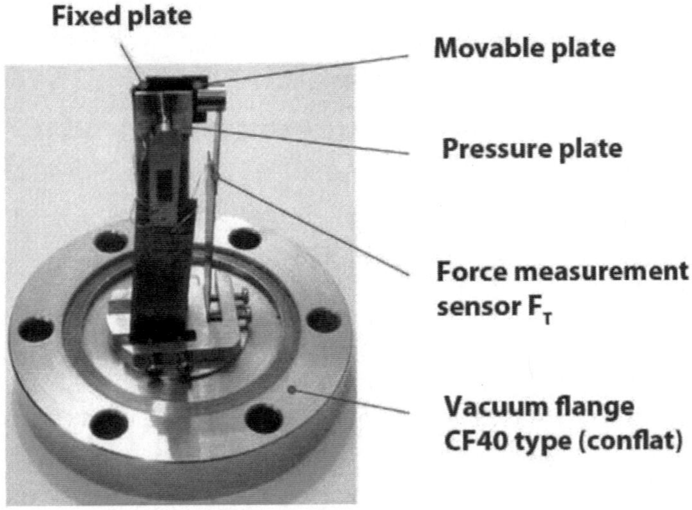

Figure 1. View of the measuring unit of the sensor mounted on the vacuum flange wherewith to measure the cleanliness of the working surface of the product in the vacuum.

The working principle of the sensor (figure 2) based on the use of breakaway force dependency F_T contact surfaces of the plates (movable, the pressure and fixed plates) of the sensor, simulating the working surface of the test part, from the ratio of the sorbate surface [4].

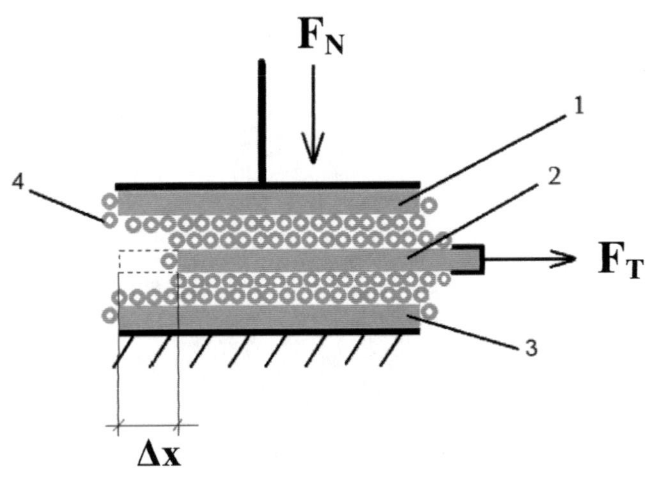

Figure 2. Principle of operation of sensor for the working surface cleanliness definition in vacuum: 1 – clamping plate; 2 – moving plate; 3 – immobile plate; 4 – adsorbed gas (sorbate).

Investigations of the static friction coefficient in over wide range of vacuum and theoretical calculations showed the possibility of a unified view of the various initial parameters, such as the residual gases pressure, temperature, relative humidity, physical properties of material etc. as the function of the surface coverage [5–10].

The measuring unit of the sensor is located in the working area of the product and determines the breakaway power F_T. Further, the measured force value F_T is supplied to the information processing unit, where, taking into account the tracking pressure F_N, material properties surface of the test part and the environmental parameters (pressure, temperature, and humidity), calculate the surface plates value according to calibration curve. Sensor features a removable plate, on which is applied a thin material cover identical to the material of the test part. Considering that the measuring unit of the sensor located close to the controlled products, when the sensor is accepted that the amount of molecular layers of adsorbed gases on the product working surface and on the contacting surfaces of the sensor plates is equal.

The calculated value of the cleanliness of the product working surface as the amount of molecular layers of adsorbed gases showed on the information processing display.

References

[1] Kemenov V N and Nesterov S B 2002 *Vacuum technique and technology* (Moscow: MEI)
[2] Krymov D V *et al* 2012 *Vacuum technique and technology* **2** 123–7
[3] Rozanov L N 2012 *Vacuum technique and technology* **4** 197–202
[4] Patent application RU 2016115692 2016
[5] Deulin E A 1998 *Book of Abstracts of IVC-14 (Birminghem)* p 310
[6] Deulin E A, Gatsenko A A and Loginov A B 1999 *Surface Science* **433-5** 288–92
[7] Deulin E A 2001 *Mechanics and Physics of Precise Vacuum Mechanisms. V. 1* (Vladimir)
[8] Deulin E A Patent RU 2316744
[9] Deulin E A 2010 *Friction & Lubrication in Machines and Mechanisms* **9** 43–7
[10] Deulin E A 2012 *Vacuum technique and technology* **3** 141–52

Journal of Physics: Conference Series **729** (2016) 012023 doi:10.1088/1742-6596/729/1/012023

Electromagnetic system for the management of the output power of the carbon dioxide laser

S A Martsinukov[1], D K Kostrin[1], V V Chernigovskiy[1] and A A Lisenkov[2]

[1] Department of electronic instruments and devices, Saint Petersburg Electrotechnical University "LETI", 197376, Saint Petersburg, Russia
[2] Institute of Problems of Mechanical Engineering, Russian Academy of Sciences, 199178, Saint Petersburg, Russia

E-mail: sergm2006@mail.ru

Abstract. The methods to control the output power of the gas-discharge lasers are shown. An electromagnetic system for the management of the output power of the carbon dioxide laser is described. The results of calculation and modeling of the magnetic field in the working gap of the electromagnetic system are presented. Experimental studies on the distribution of magnetic induction in the electromagnetic system are carried out.

Laser technology is used in various technological fields. From all available types of lasers of greatest interest is a gas-discharge carbon dioxide laser (CO_2 laser) which can be used for cutting glass and welding of metals, and also for medical operations in cosmetic surgery and dentistry. Carbon dioxide lasers can have very high output power – up to tens of kW in continuous mode. The advantages of pulse mode operation of the technological CO_2 laser are due to the nonstationarity nature of the pumping and generating processes. Technological operations require both smooth change of output power and the formation of a pulse with the specified amplitude and shape with a temporal resolution of about 1 ms.

Existing methods of the CO_2 laser power management are not always optimal, which greatly complicates their application. Power modulation can be carried out using [1]:

– mechanical modulators. They are simple and reliable, but do not allow to smoothly change shape, duration and average power of a laser pulse;

– electro-optic modulators. They have low inertia, but have a strong sensitivity to damage and low radiation resistance of the crystal, also they are very expensive;

– acousto-optic modulators. These modulators have low radiation resistance of the crystal and a very large inertia;

– current (internal) modulators. These modulators are quite common; they do not require additional optical elements, which simplifies the overall laser system. However, under a significant change in current of the laser occurs oscillatory process of establishing the coherent laser power caused by temperature change of the resonator and a change in the composition of the gas mixture.

To eliminate these disadvantages it is advisable to use an electromagnetic modulation. In this case, for modulation of the CO_2 laser power a transverse magnetic field is used [2, 3]. Electromagnetic modulation method is based on the plasma-optical effect, which is due to the interaction of the magnetic field with the active medium of the laser. This method allows controlling the output power of the laser by changing its value from maximum to zero according to a linear law with a sufficient speed.

In the sealed-off gas-discharge lasers of medium pressure action of a transverse magnetic field on the output power is due to changes in the zone of interaction of the active medium with the field of the resonator and discharge current – with increase of the magnetic field induction these values decrease. The performance of the magnetic control of the CO_2 laser is determined by the time of plasma movement inside the tube, the rise time of the magnetic field, the lifetime of excited molecules and the settling time of the radiation in the resonator.

For the CO_2 laser controlled by a transverse magnetic field in dynamic mode, the time lag of the output laser power relative to the input signal (current through the control coil when a magnetic field is generated by a system of electromagnets) depends on the following main factors:

– delay of the increase of the magnetic field;

– end-time of the discharge movement in a magnetic field.

Since rigorous theoretical analysis of the developed electromagnetic system is very complex the most appropriate way is its modeling using special programs and experimental study of the prototype.

For the magnetic control of the discharge it is necessary to determine the geometry of the magnetic circuit and to calculate the parameters of the circuit for the coils of the electromagnets. Changing the geometry of the pole pieces it is possible to change the cross-sectional area of the working gap in the area of management. Increasing this area reduces the average magnetic induction in the control gap, but the magnetic field becomes more uniform and decreases the share of the field, lost due to scattering.

Calculation of the magnetic field distribution in the working gap was performed using the software package FlexPDE [4]. FlexPDE is a software program designed to build scenario models for the solution of differential equations by the finite elements method. FlexPDE produces operations required to convert the description of the system of differential equations into the model for calculation by the finite elements method, finds solution for this system and presents the results in a graphical form.

The first stage of modeling was to estimate the homogeneity of the magnetic field in the working gap when using from one to seven electromagnets (the limit on the maximum number of electromagnets equal to seven is due to the dimensions of the laser tube). Coils of the electromagnets intended for use in the developed system have a frame made of organic glass with a thickness of 5 mm to prevent breakdown, which is wound with insulated wire with a diameter of 1.5 mm. Coil consists of 9 layers and 23 turns in each layer. Figure 1 shows plots of the distribution of the magnetic field B along the tube of the laser depending on the position x relative to its center for a system of one, three and seven electromagnets.

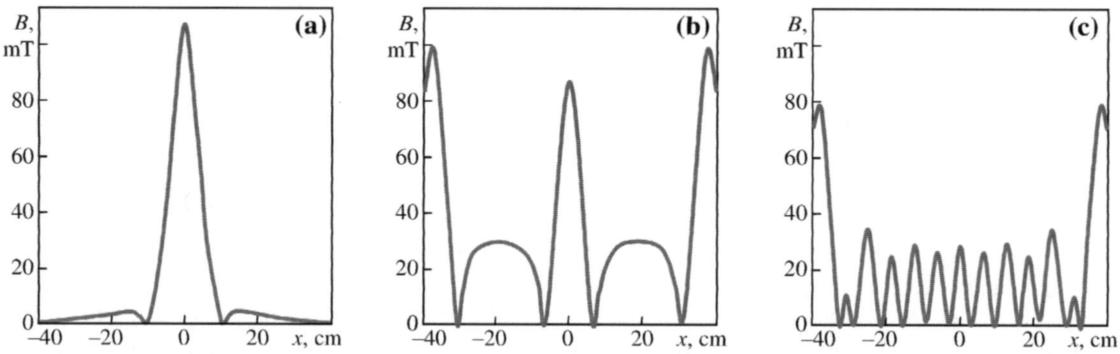

Figure 1. Plots of the distribution of the magnetic field B along the tube of the laser depending on the position x relative to its center for a system of one (a), three (b) and seven (c) electromagnets.

As the number of magnets increases the uniformity of the magnetic field along the tube of the laser improves, this reduces the maximum amplitude of the induction. It is obvious that the uniformity of the magnetic field even when using seven electromagnets (figure 1(c)) is not satisfactory and this system is not suitable to control the output laser power.

To correct the magnetic field inhomogeneity can be used a special equalizer that allows aligning the value of the magnetic induction along the laser tube [5, 6]. Figure 2 shows charts of the magnetic field induction distribution along the laser tube, depending on position relative to its center for the system of equalizers and one, three and seven electromagnets.

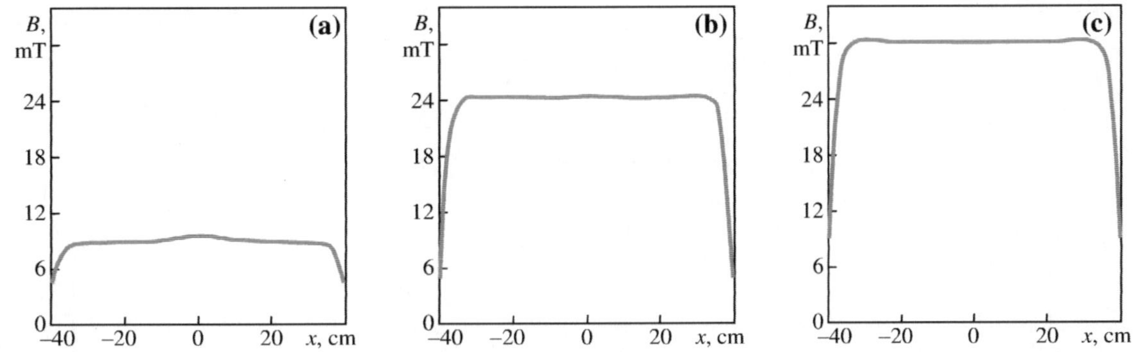

Figure 2. Charts of the magnetic field induction distribution along the laser tube, depending on position relative to its center for the system of equalizers and one (a), three (b) and seven (c) electromagnets.

The use of the equalizer allows aligning the magnetic field even when using a single electromagnet (figure 2(a)). However, with increasing number of magnets the alignment of the field distribution along the laser tube improves and the value of the magnetic induction increases. The program FlexPDE allows not only to calculate the distribution of the magnetic field, but also to get the pattern of the magnetic field distribution in the system being developed. Results of the calculation of the magnetic field distribution in the central part of the working gap and at the edge of the electromagnetic system are shown in figure 3.

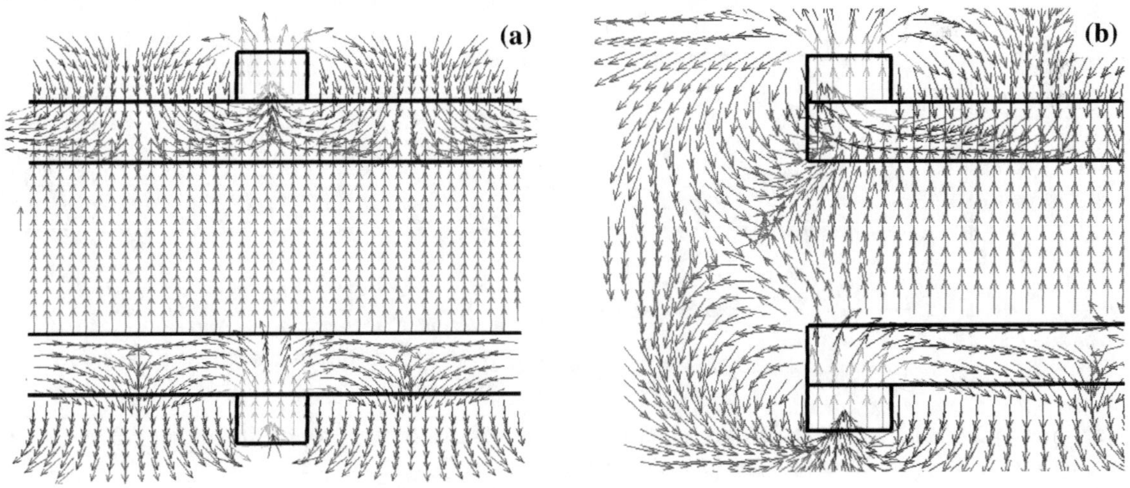

Figure 3. Results of the calculation of the magnetic field distribution in the central part of the working gap (a) and at the edge of the electromagnetic system (b).

Simulation results gave an opportunity to design and assemble prototype of the electromagnetic control system for the carbon dioxide laser. Measurements of the magnetic induction in the gap of the developed system was carried out with the help of milliteslameter. The measuring procedure was as follows: the space between the equalizers was divided into the cells of 10×10 mm and at the point of intersection of the grid lines was carried out the measurement of the magnetic field. The measurements were carried out for both longitudinal and transverse components of the magnetic induction. Figure 4(a) shows the dependence of the magnetic induction in horizontal planes when the current of the electromagnets I = 5 A at different distances y from the center of the system in the transverse plane. Figure 4(b) shows the distribution of the transverse component of the magnetic field with the same current of the electromagnets at various distances x from the center of the system in the longitudinal plane.

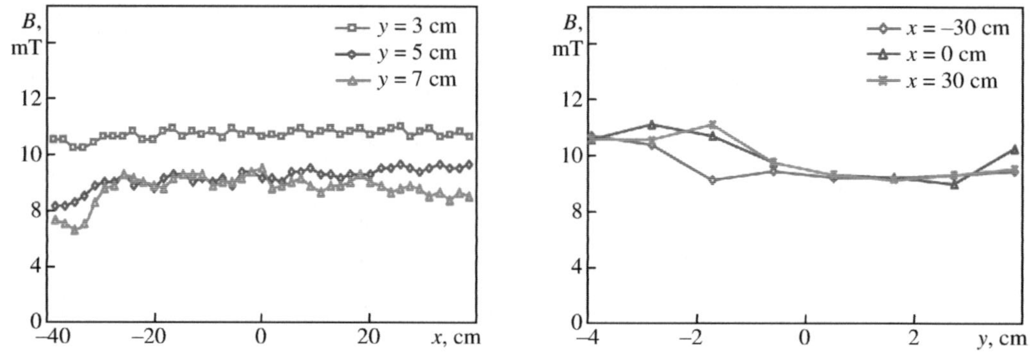

Figure 4. Dependence of the magnetic induction in horizontal planes when the current of the electromagnets I = 5 A at different distances y from the center of the system in the transverse plane (a) and the distribution of the transverse component of the magnetic field with the same current of the electromagnets at various distances x from the center of the system in the longitudinal plane (b).

Based on the above graphs, we can conclude about the uniformity of the magnetic field distribution in the developed system, both in longitudinal and transverse section. Of additional interest is the investigation of the dependence of the magnetic field from the current through the coils of the electromagnets. Figure 5 shows the dependence of the magnetic induction in horizontal planes at different currents of the electromagnets.

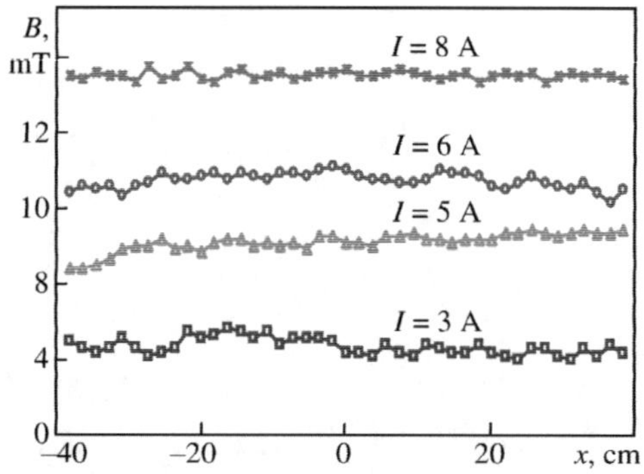

Figure 5. Dependence of the magnetic induction in horizontal planes at different currents of the electromagnets.

According to the study of the developed electromagnetic system for control of the output power of the carbon dioxide laser we can draw the following conclusions:

– with the increase of operating current of the coils of the electromagnets the magnetic field is increasing and has almost linear character;

– range of the values of the magnetic induction decreases with the increase of operating current of the coils of the electromagnets;

– distribution of the magnetic induction along the laser tube is almost uniform.

Journal of Physics: Conference Series **729** (2016) 012023 doi:10.1088/1742-6596/729/1/012023

Thus, the developed control system could allow not only to efficiently manage the average power of the CO_2 laser, but also to acquire pulses of radiation while using laser designed for continuous operation.

References

[1] Elokhin V A and Zhdanov I G 2003 *Scientific Instrumentation* **3** 46–51
[2] Martsinukov S A and Chernigovskiy V V 2004 *Proceedings of SPbETU "LETI"* **2** 27–31
[3] Martsinukov S A 2013 *Proceedings of SPbETU "LETI"* **4** 13–7
[4] Martsinukov S A and Kostrin D K 2014 *Proceedings of the Russian Universities: Radioelectronics* **1** 50–4
[5] Chernigovskiy V V and Martsinukov S A 2013 Patent RU 125780
[6] Chernigovskiy V V and Martsinukov S A 2013 Patent RU 126206

Journal of Physics: Conference Series **729** (2016) 012024 doi:10.1088/1742-6596/729/1/012024

Mass transfer in ablation process with large angle of laser ray incidence on target and small distance between target and substrate

A A Lozovan[1], S V Prishepov[1], S V Frangulov[1], S S Aleksandrova[1], R N Rizakhanov[2] and S K Sigalayev[2]

[1] Moscow Aviation Institute (National Research University), ul. Orshanskaya 3, Moscow, 121552, Russia
[2] Keldysh Research Center, ul. Onegskaya 8, Moscow, 125438, Russia

E-mail: loz-plasma@yandex.ru

Abstract. This paper presents the results of research of laser ablation, carried out at 85° incidence angle of the laser ray to the normal to surface of target with simultaneous spatial restriction of plasma torch. It is shown that laser radiation reflected from the target falls on the substrate and produces ablation. Consequently ablated material of the substrate is transferred to the target. It is found, that direct and reflected from the target laser radiation form periodic wave-shaped structures on the surface of target and substrate.

1. Introduction

Wide range of possibilities of the method of pulsed laser deposition (PLD) makes it a subject of a growing interest in the industry [1]. One of the unique features is the ability of laser radiation to generate plasma at a considerable distance from the laser and in the inner cavities of various products [2], that in the end allows to modify internal surfaces of cavities, including coating by PLD method. Characteristics and technological capabilities of spatially bounded plasma formed by laser radiation in the quasi-closed volume of hollow bodies now is not fully understood, despite the theoretical and experimental studies held by different groups of scientists [3]. With spatial restrictions significantly increases the concentration of excited particles in the laser plasma, what proved very useful in laser-induced breakdown spectroscopy [4], and also changes the shape of the plasma torch. When plasma torch is directed forward (obvious characteristic of the laser-induced plasma) its parameters and shape substantially depend on target-substrate distance. When the distance is small the interaction of the substrate and plasma torch should significantly affect morphology of the film deposited on the surface. Most researches of laser ablation and PLD were carried out at $\theta = 45°$ incidence angle of the laser ray to the normal to surface of the target. There are still few researches at $\theta \geq 80°$, although they are of considerable interest for a variety of practical applications: analysis of the chemical composition of the target material [5], development of efficient soft X-ray lasers and plasma accelerators [6], micromachining [7], medicine [8], etc. Ablation with $\theta \geq 80°$ and spatial restriction of plasma during film deposition process was used in [9]. However, the experiments were carried out at atmospheric pressure using the method of air blowing for removing particulates from the film deposition zone, which is unacceptable for the deposition on the internal surfaces of cavities. Previously, during experiments with film deposition by PLD method with small target-substrate distance we have found

Journal of Physics: Conference Series **729** (2016) 012024 doi:10.1088/1742-6596/729/1/012024

re-evaporation of the film by laser radiation reflected from the target [10]. The experiments were performed in studies on the coating of internal surfaces of pipes at angles of incidence of laser radiation at the target 20°, 30° and 60° to the normal to the target surface. In this paper ablation was performed at 85° angle of incidence to the normal to surface of target, i. e. close to sliding angle, with simultaneous spatial restriction of plasma torch. Results of this research are interesting in the aspect of studying of modification of inner surfaces of hollow bodies.

2. Experimental

Experiments were conducted on laser stand [2]. Ablation was made by YAG:Nd laser LTI-215(M) with pulsed pumping and modulated quality factor. Parameters of laser radiation: wave length λ = 1064 nm with maximum pulse energy I = 180 mJ; pulse length τ = 10 ns; pulse frequency f = 30 Hz; beam divergence γ = 5 mrad. Laser radiation was focused on the target by a lens with focal length F = 100 mm. The size of laser mark on the target was 1 mm, therefore the power density of laser radiation on the target was $W = 2.2 \cdot 10^9$ W/cm^2. A more detailed description of the installation can be found in [11]. With an incidence angle 85° to the normal, the laser mark on the target takes the form of an extremely elongated ellipse 7 mm in length, so its area increases 6 times in comparison to one formed at the normal incidence, and the power density of laser radiation on the target is correspondingly reduced to $3 \cdot 10^8$ W/cm^2. A diagram of the experiment is shown in figure 1. Plates with 1 mm thickness were used for both the target and substrate. Material of the target: alloy Ti (~Al = 7 %, ~Si = 0.5 %), and substrate material is polished steel SUS 304 of industrial supply. Number of pulses – 150.

Target was set at 3 mm distance from the center of laser beam mark, thus providing significant plasma torch restriction. Target and substrate surfaces were chemically cleaned and then placed in the working chamber, which was vacuumed to pressure $P = 10^{-3}$ Pa. Ablation was conducted at pressures 10^{-3} and 50 Pa, under Ar presence. Morphology of surface and chemical analysis of specimens were studied by means of optical microscopy on microscope "Carl Zeiss Jena" INTERFACO A_E and scanning electron microscope FEI Quanta 600FEG with microanalysis system EDAX Trident XM4.

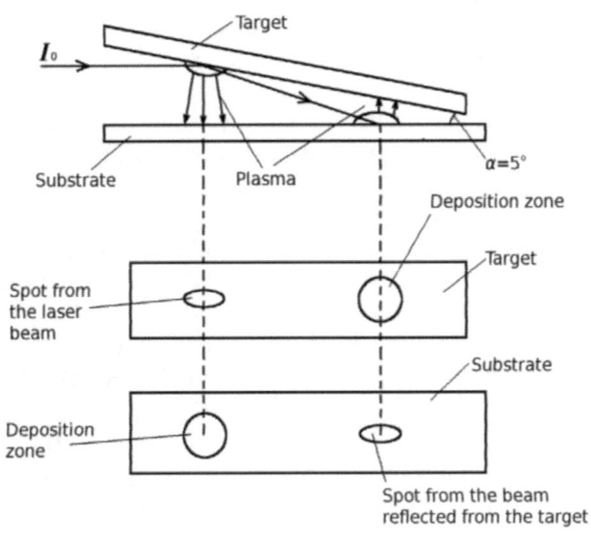

Figure 1. Experiment scheme.

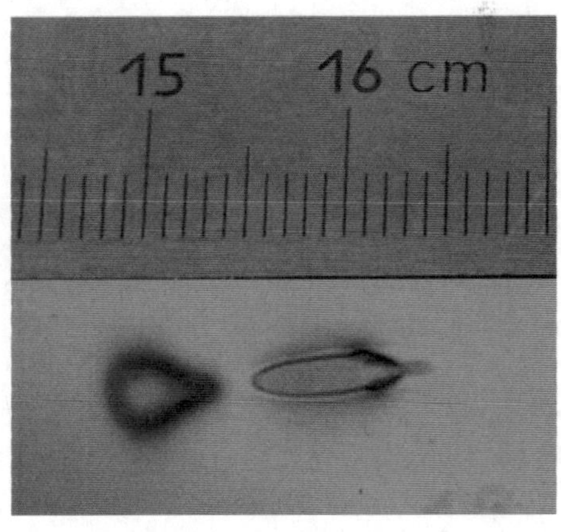

Figure 2. Traces on the substrate.

3. Results and discussion

As a result of ablation material of the target transferring to the substrate (figure 2), the spot of the coating is formed in an almost-round shape, in spite of the elliptical shape of the laser radiation spot on the target. This happens because of incomplete «flip-over effect», which is theoretically described in [12]. This effect results in a 90° rotation of the ellipse on the substrate relative to the ellipse on the target, due to the higher pressure of torch vapors dissipating in a radial direction, relative to the direction to the surface of the target. In this case, extremely small distance between target and substrate does not allow the «flip-over» process to fully complete. Microanalysis of elements in the center of the film deposition zone found an increase in Ti up to 6.21 % in comparison to its content in steel substrate 0.4 % after ablation by 150 pulses under high vacuum. This confirms efficiency of ablations at a high angle of incidence of the laser ray to the target. At the same time during the ablation by 10 pulses, the content of Ti is significantly lower. In our opinion this is due not only to the difference in the total duration of radiation in both modes, but also to the stronger development of relief of the target surface under 150 pulses, which increases the efficiency of ablation.

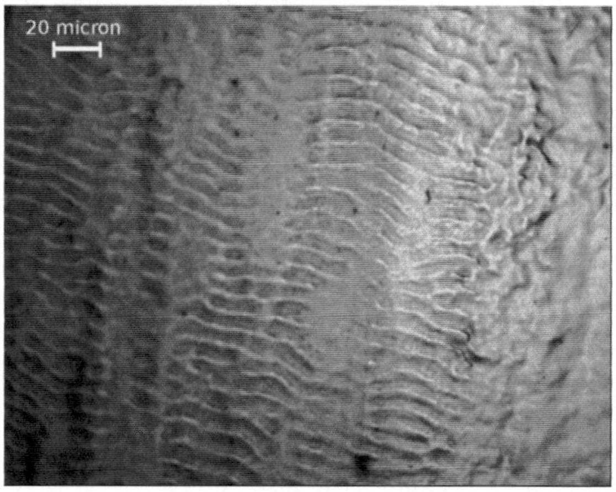

Figure 3. Laser mark on the Ti target (150 pulses, $P = 10^{-3}$ Pa).

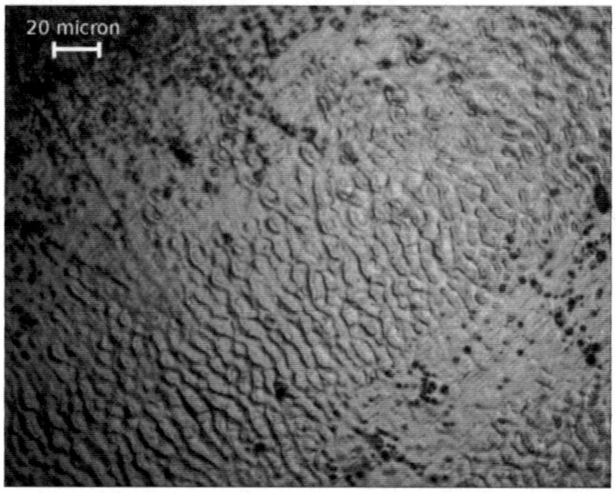

Figure 4. Laser mark on the Ti target (10 pulses, $P = 10^{-3}$ Pa).

Photographs of the centre of the laser radiation mark on the surface of Ti target after 150 and 10 laser pulses and pressure $P = 10^{-3}$ Pa are shown in figures 3 and 4. Both pictures show structures of periodic waves. Short elements of wave formed at 10 pulses, and then grow together in long waves at 150 pulses. Thus, the period of the wave stays the same. At 50 Pa pressure periodic structures still exist, but their length stays short at 150 and 10 laser pulses. Periodic wave structure-forming-processes are initiated by laser irradiation of surfaces of semiconductors, metals, polymers, dielectrics and fluids at different parameters of laser ray and are detailed in different literature sources.

At oblique fall of laser ray at large angle to the normal to the surface, strong reflection of radiation from the surface is going on along with ablation. Reflected ray at small target-substrate distance falls to the substrate and ablates its surface. Ablated material transfers to the surface of the target, i. e. on this stage of the process target becomes substrate. Reflected from the target ray makes mark on the processing surface also in the shape of strongly elongates ellipse with the same length approximately 7 mm (figure 2).

Photo of the surface of substrate in the center of the reflected laser radiation impact zone is given in the figure 5. Highly developed wavy surface is shown, and that says about surface melting under irradiation and accordingly about significant value of reflected radiation. Thus, influence picture of reflected radiation on the substrate surface is similar to the influence picture of direct laser radiation on the target with only difference in length of the wave elements, which in this case are shorter.

Figure 6 presents the photo of zone of film deposition spot on the target, produced by the ablation of substrate by reflected ray. We see cracks formed on the film. Spectra of the target and of this film are shown in figures 7 and 8. It is seen that significant transfer of material from substrate to target has occurred.

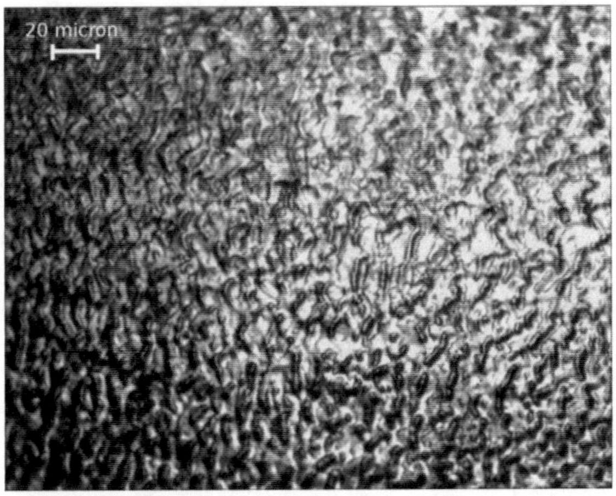

Figure 5. Zone of reflected radiation action on the substrate after 150 pulses ($P = 10^{-3}$ Pa).

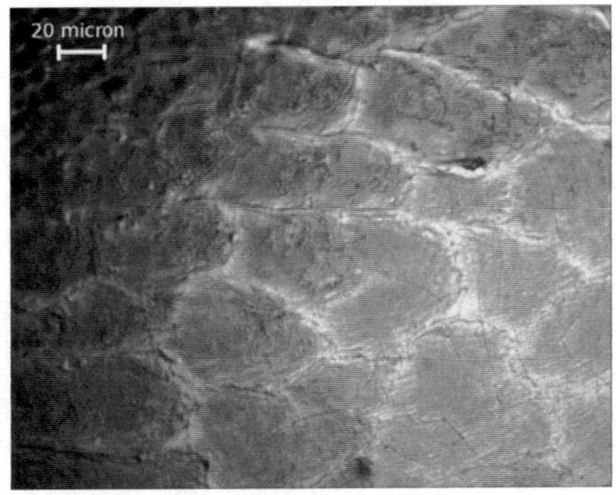

Figure 6. Zone of the film on the target, produced by reflected radiation after 150 pulses ($P = 10^{-3}$ Pa).

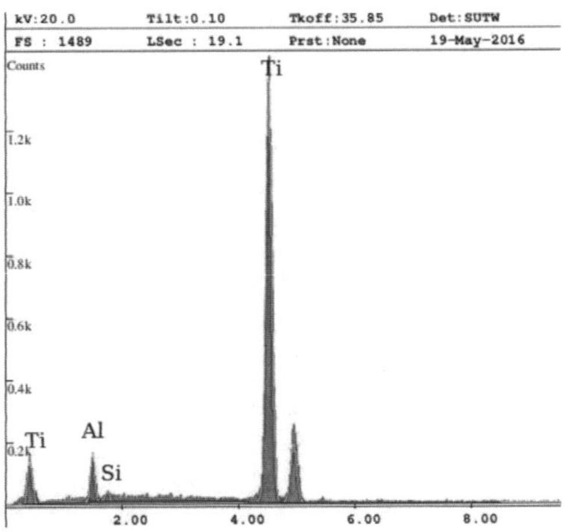

Figure 7. Spectrum of target before irradiation.

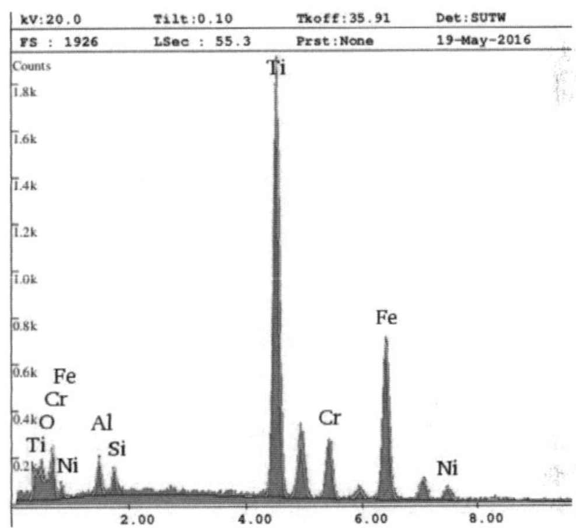

Figure 8. Spectrum of deposition spot zone on Ti target, produced by ablation by the ray reflected from the substrate at 150 laser pulses ($P = 10^{-3}$ Pa).

4. Conclusions

Research of laser ablation of Ti alloy at laser radiation angle of incidence on the target 85° from the normal to the surface with simultaneous spatial restriction of plasma torch are carried out. It is shown, that in the result of significant reflection of laser radiation from the target part of it falls on the substrate and produces ablation. Ablated material of the substrate transfers to the target. It is determined that direct and reflected from the target laser radiation forms periodic structures of waves on the surface of target and substrate consequently.

Acknowledgments

This study was supported by the Ministry of Education and Science of the Russian Federation within the scope of the state task for research efforts (project code 3087).

References

[1] Chrisey D B and Hubler G K 2016 *Pulsed Laser Deposition of Thin Films*
[2] Lozovan A 2003 *Vacuum* **70** 477
[3] Gao X, Liu L, Song C and Lin J 2015 *J. Phys. D: Appl. Phys.* **48** 175205
[4] Hou Z, Wang Z, Liu J, Ni W and Li Z 2014 *Opt. Express* **11** 12909
[5] Uebbing J, Ciocan A and Niemax K 1992 *Spectrochim. Acta B* **47** 601
[6] Wang Y, Luther B M, Berrill M, Marconi M, Brizuela F and Rocca J J 2005 *Phys. Rev. E* **72** 026413
[7] Ahmmed K M T, Grambow C and Kietzig A-M 2014 *Micromachines* **5** 1219
[8] Arba-Mosquera S and Verma S 2013 *Biomed. Optics Express* **8** 1422
[9] Konov V I, Kononenko T V, Loubnin E N, Dausinger F and Breitling D 2004 *Appl. Phys. A* **79** 931
[10] Lozovan A A, Alexandrova S S, Mishnev M A and Prishepov S V 2014 *J. of Alloys and Compounds* **586(1)** S387
[11] Lozovan A A, Frangulov S V and Chulkov D V 2011 *Laser Beams: Theory, Properties and Applications* (New York: Nova Science Publishers Inc.) chapter 8 pp 331–49
[12] Anisimov S I, Luk'yanchuk B S and Luches A 1996 *Appl. Surf. Sci.* **96–8** 24

Consideration of the condensation processes of thin films in the crystal substrate's potential field

V A Tupik, V I Margolin and Chu Trong Su

Department of Radio Microelectronics and Radio Equipment Technology, Electrotechnical University "LETI", Saint Petersburg, 197376, Russia

E-mail: chusu171@mail.ru

Abstract. The condensation process of a single particle in an ideal crystal substrate's potential field is considered. The optimal deposition path and the potential barrier of deposited particle's motion are shown. Some computer modeling examples of thin film's growth process were carried out on the basis of the implemented programs. A fractal analysis of obtained thin films was made, on the basis of which the possibility of estimating the performance of thin film's growth process will be discussed.

1. Introduction

Obtaining thin films of high quality today is constantly required, especially for micro-nanoelectronics industries and the formation of various functional coatings. A major problem in the way of improving the quality of thin films has to do with the analysis of the conditions and the processing of how the films are obtained. In this paper, the process of film condensation will be examined as one of the possible ways of obtaining high quality thin films.

2. Assumption

The theories adopted in this work, are as follows:

a) material consists of spherical particles (atoms, molecules), having isotropic interactions with other particles through the Lenard–Jones pair potentials (LJ) [1]. This approach characterizes a simple model calculation but has some limitations [2];

b) substrate has an ideal structure and does not change with time of calculation.

3. A procedural analysis of the process

The analysis of the process is done by three successive calculation models, which are as follows.

3.1. Condensation process of single particle on substrate

In considering the condensation process the firstly we want to examine the process for a single particle on the potential field of an ideal crystal (without taking into account the deposited particle's kinetic energy). The potential bond $U(x, y, z)$ of deposited particles with substrate was calculated by the following equation:

$$U = U_{\text{1-layer}} + U_{\text{2-layer}} + U_{\text{3-layer}} + \dots, \qquad (1)$$

where

$$U_{n\text{-layer}}(x,y)=\sum_{l=-8}^{9}\sum_{k=-8}^{9}\left[\frac{V_{ij}}{[[[(l-b)\cdot y_0+\Delta y]^2+[(k-a)\cdot x_0+\Delta x]^2+z_n^2]/б_{ij}]^6}-\right.$$
$$\left.-\frac{V_{ij}}{[[[(l-b)\cdot y_0+\Delta y]^2+[(k-a)\cdot x_0+\Delta x]^2+z_n^2]/б_{ij}]^3}\right], \quad (2)$$

and $V_{ij}=\sqrt{V_iV_j}$; $б_{ij}=(б_i+б_j)/2$; $\Delta x=x-x_0$; $\Delta y=y-y_0$; $\Delta z=z-z_0$; $a=b=0$, if n – is an even number, otherwise $a=b=0.5$; $V_{ij}, б_{ij}$ – Lenard–Jones parameters for interactions occuring within materials (particles) i and j, respectively; n-layer – the number of n layer of ideal substrate (in this paper n = 9, the substrate has parameters of centred unit cells: edge lengths x_0, y_0, z_0 and axial angles $\alpha=\beta=\gamma=90°$). Using this computed model in figures 1 and 2 results shown in the example used the following parameters $V=1$, $\sigma=1$, $x_0=y_0=z_0=r_{opt}=2^{1/6}$(where r_{opt} – the distance at which the normalized pair Lenard–Jones potential was minimum). For convenience, the process of modeling all the parameters of the potential energy and the length will have standardized values as well as graphs dimension. For the potential energy, the standardized value $V_0=10^{-21}$J and corresponds to the real value $V_{real}=V\cdot V_0$; $\sigma_0=10^{-10}$m and $r_{real}=r\cdot\sigma_0$ for the length, respectively. Based on the calculated results the optimal path of the deposited particle on a substrate is represented by the curves $x(z)$ and $y(z)$ in figure 1. The potential interaction of deposited particle with ideal substrate was realized as indicated by the minimum (curve $U(z)$ in figure 1). Evident was the fact that the potential surface relief on the substrate changes, and the presence of a potential barrier in deposited particle's motion on the substrate (figure 3). These factors will be confirmed below in computer modeling of thin film growth process taking into account the thermal noise, and excluding the thermal noise on the substrate surface (figure 4).

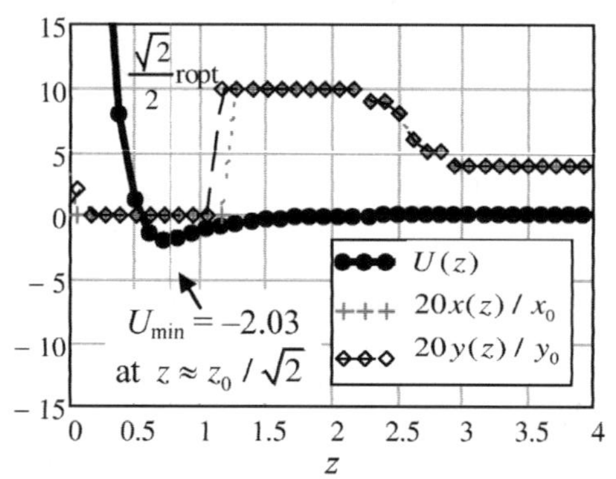

Figure 1. The optimal condensation path $x(z)$ and $y(z)$, and the potential interaction of deposited particle with ideal substrate.

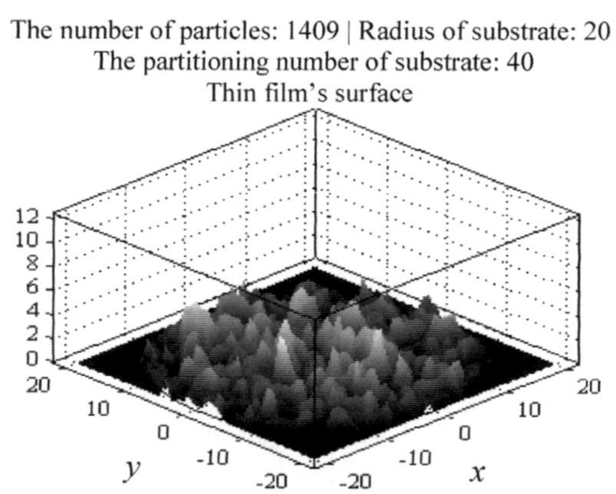

Figure 2. Simulation of the roughness of the deposited thin film.

3.2. Computer modelling of thin film growth process

In order to visually consider the process of condensation and growth of thin films on potential field substrate was built a designated program [3], the foundations of which are based on the Monte Carlo method, to select a number of particles and quasi-Newton lattice's method in determining the optimal direction of motion of the particles, such that the potential energy of the system (consist of deposited particles) was minimal. As you know, this process is random (using the Monte Carlo method) and the self-organizing nature of the (using quasi-Newton lattice's method) with a phase transition [4].

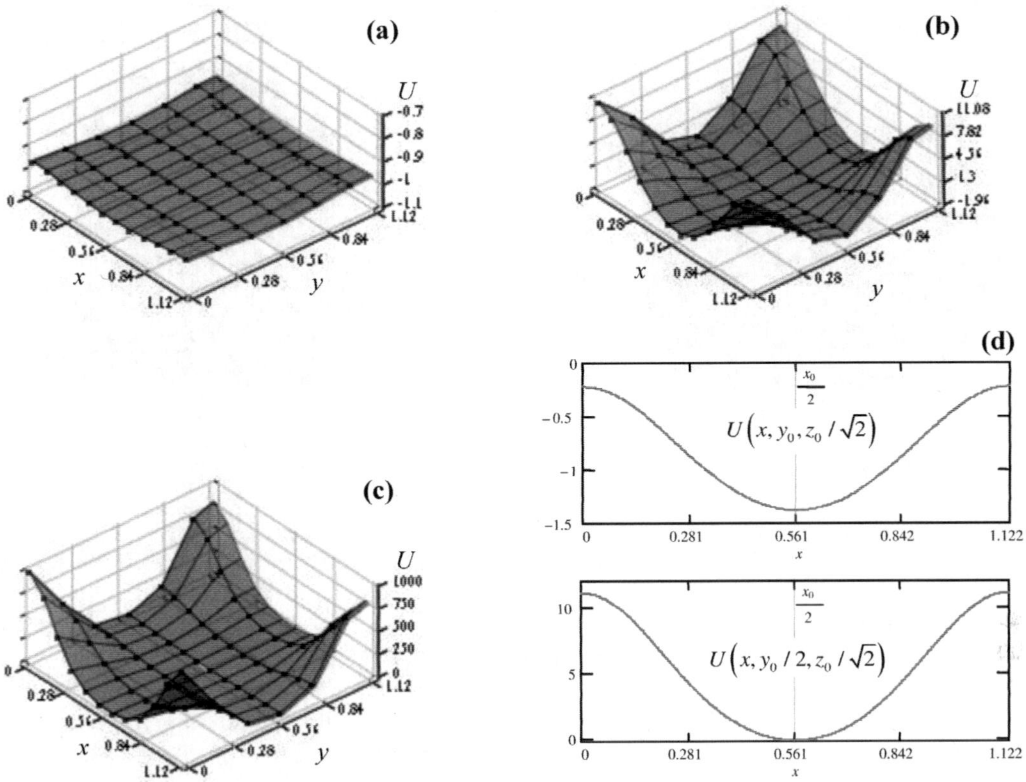

Figure 3. Topology relief: (a) – $U(x, y, z = z_0)$; (b) – $U(x, y, z = z_0/\sqrt{2})$; (c) – $U(x, y, z = z_0/2)$. Profile potential relief in two cases (d).

Below are shown the results of computer modeling for such data, 1409 particles randomly generated and successively dropped onto a substrate (the structure and properties of the particles were already set up in the calculation model A). This is the number of particles selected as sufficient to cover the substrate with a radius of 20; precipitated by lowering the particle flow rate chosen so that during lowering of a particle on a substrate is realized 500 + 8N steps traffic (N – number of particles in a real-time simulation) in lattice grid calculating coordinates (the step-size selected is $r_{opt}/10$ – the lesser the step-size the more precise the calculation, but the calculation time is longer). Figure 4 shows the result of computer simulation of the particle deposition process parameters σ and V are same as the parameters of the particles on the substrate, the left takes into consideration thermal noise while the right excludes it, and the noise level is from 10^{-1} to V_0. It can be seen that the thermal noise allows the particles to remove a potential barrier (which we calculated in the calculation model A) and formed larger clusters; however the number of defects (disorder) in the picture is more.

3.3. Fractal analysis for obtained thin film

The basis of fractal analysis of obtained thin film structure; as is known, the fractal dimension of the physical objects may take on different values depending on the method of calculation and for different regions (volume surface or section) may be different (result confirmed in table 1 and figure 5). The fractal dimension of the resulting thin film was calculated in this paper using three known methods [5, 6].

The first method: box counting method, (Hausdorff–Besicovitch) fractal dimension Db of obtained structures was estimated by the formula: $N(\delta) \sim \delta^{Db}$, where $N(\delta)$ – the number of filling elementary cubes with the number of partitions of calculation area (see figure 3) (for three directions x, y, z) δ.

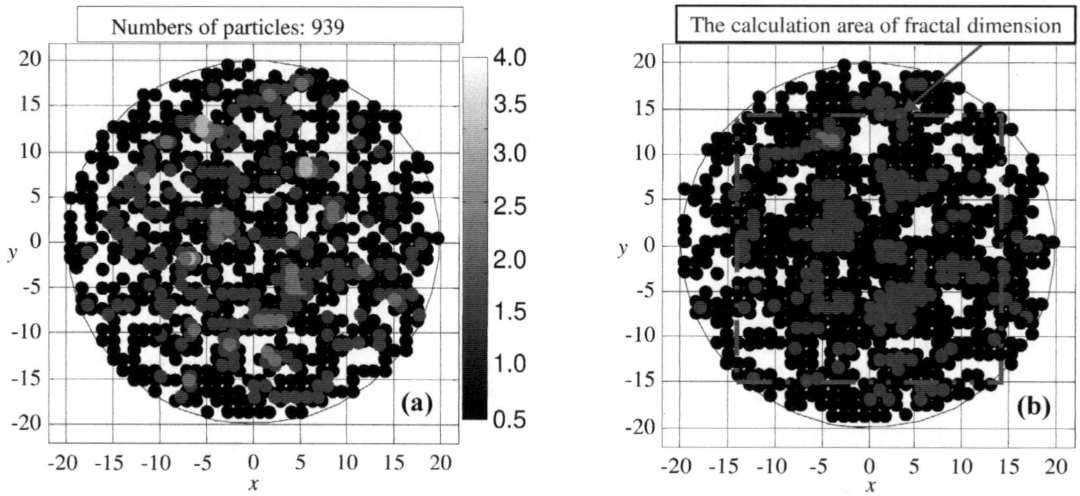

Figure 4. Computer simulation of thin film growth process without thermal noise (a) and with thermal noise (b).

The second method: fractal analysis of the surface, the surface fractal dimension D_f: $S_{fact}(\delta) \sim \delta^{Df}$, where $S_{fact}(\delta)$ – the surface area of the resulting thin film when the number of partitions of calculation area (for two directions x, y) δ.

The third method: fractal analysis of the vertical section, the fractal dimension (profile) is calculated using the formula: $L(\delta) \sim \delta^{DL}$, where $L(\delta)$ – the length of the profile of a cross section through the center of the resulting thin film with the number of its partitions (one direction) δ.

Below in table 1 and figure 4 are shown the result of calculating the fractal dimension of the resulting thin film in the scope of investigation (figure 3) at different moment in a thin film growth process, and for the two cases when the length of the Lenard–Jones parameters of the deposited particles (σ_{dep}) and particles on a substrate (σ_{sub}) are equal (epitaxial growth, figure 3(a)) and different (σ_{dep}:σ_{sub} = 2:1) According to the calculation results the fact (statistical law) that fractal dimension is increased with the thin film growth process for the three calculation methods. The longer and more complex the growth process, the greater the value of the fractal dimension.

Table 1. Result of fractal dimension's calculate for three different methods.

Fractal dimension	D_b		D_f		D_L	
Ratio of σ_{dep}:σ_{sub}	1:1	2:1	1:1	2:1	1:1	2:1
Thin film consisting of 1409 particles	1.61	–	0.58	–	0.40	–
Thin film consisting of 939 particles	1.48	–	0.36	–	0.23	–
Thin film consisting of 705 particles	1.43	1.69	0.27	1.05	0.16	0.81
Thin film consisting of 282 particles	1.10	1.30	0.11	0.47	0.06	0.31

4. Conclusion

This paper presents the result of studying the condensation process of thin films in the potential field of an ideal structure of the substrate. By considering the model of the condensation process of single particle it was shown that motion path of a condensed particle is complex, interaction potential's topography changes, and the presence of potential barrier in particle's motion on the substrate.

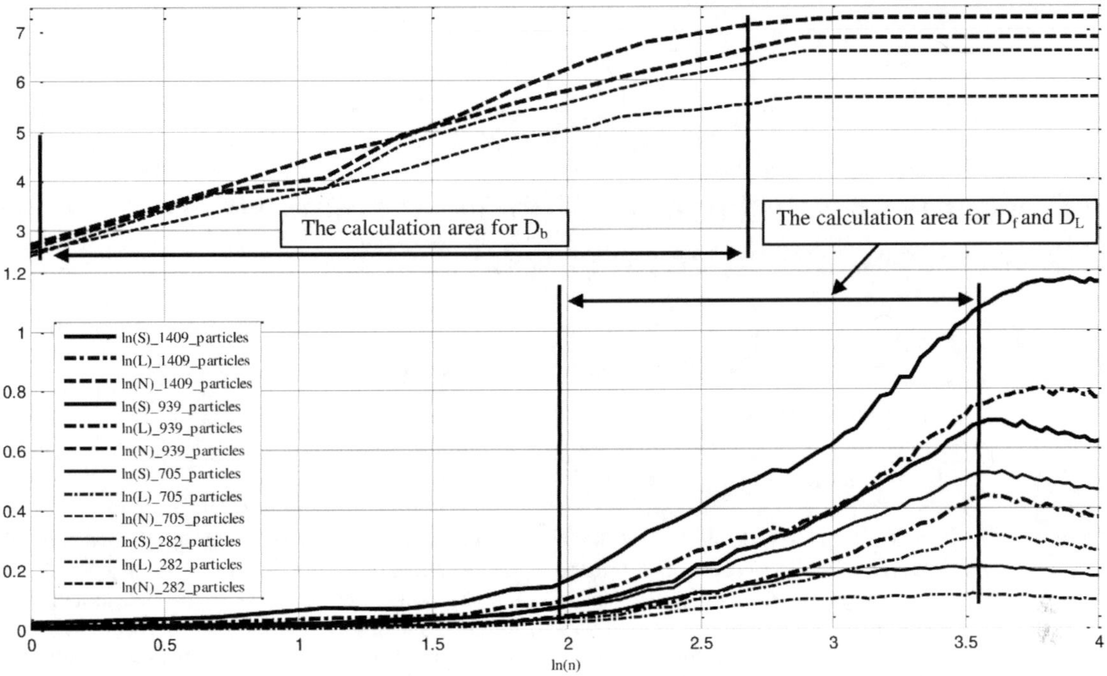

Figure 5. The dependences of ln(S), ln(L), ln(N) on ln(n) for obtained thin films in different moments of its growth process.

Thanks to the efficiency of the computer simulation of thin films growth process these facts have been confirmed, and it is also shown that the potential barrier can be removed with the thermal noise (thermal treatment). According to the results of fractal analysis of the obtained thin film structure, it can be concluded that this thin film has a fractal structure and moreover, by the value of fractal dimension the difficulty and time of their growth process can be determined.

References

[1] Rieth M 2003 *Nano-Engineering in Science and Technology: An Introduction to the World of Nano-Design* (New Jersey: World Scientific)

[2] Nazarov A A and Mulyukov R R 2010 *Atomistic modeling of materials, nanostructures and processes of nanotechnology* (Ufa: Bashkir State University)

[3] Margolin V I and Chu Trong Su 2015 *Computer simulation of thin films growth process in potential field (Thin film growth)* Public registration of computer program 2015610052

[4] Kukuskin S A and Osipov A V 1998 *Advances in physical sciences* **10** 1083–116

[5] Lopes R and Betrouni N 2009 *Medical image analysis* **13** 634–49

[6] Torkhov N A, Bozhkov V G, Ivonin I V and Novikov V A 2009 *Semiconductors* **1** 38–47

Tribochemical peculiarities of lubricant composition with surface-modified metal powder

A G Syrkov, M O Silivanov and A N Kushchenko

Mining University, St. Petersburg, Russia

E-mail: syrkovandrey@spmi.ru

Abstract. The influence of different factors (adhesion, surface hydrophobicity, etc.) on antifrictional properties of lubricant with modified metal additives was discussed. The measurings of friction coefficient (f) and friction force (F_{fr}) were carried out for the heterogeneous systems as oil I-20 with Al-additives modified by triamon (T), alkamon (A) and ethylhydridesiloxane according to various programs. It was established that as a number of T-underlayers, included in Al-additives with chemisorpted external layer of ethylhydridsiloxane reduces from 3 to 1 the force of friction and coefficient of friction reduce. It was discovered that the value of summand which stands for the amount of intermolecular forces in the boundary friction equation can be regulated in Al-additives by using low-molecular T-underlayer.

1. Introduction

Consistent (mixed) chemisorption of quaternary ammonium compounds (triamon and alkamon, which are surfactants, based on quaternary ammonium compounds, with molecules having different size) proved to be a potentially successful method of regulation of water repellency, reaction ability and anti-frictional properties of modified metals [1, 2]. The most interesting research objects are modified aluminum samples and Al-containing lubricants. This due to properties listed above and the synergetic effects observed [1–3]. Hydrophobization of metal additives favorably affect anti-frictional properties of lubricants [1, 4], however the influence of location and number of underlayers on metal, with external hydrophobic layer, on adsorption properties of that metal and on characteristics of lubricants is still unknown. Also, fundamental tribological characteristic (F_{fr} and f) of systems with lubricants, which contain disperse metals with multiple layers (n = 1–4) of quaternary ammonium compounds and hydrophobic ethylhydridsiloxane, were not yet measured and described. This work's aim was to solve the problems listed above.

2. Materials and methods

Aluminum powder (PAP-2) with specific surface area of 2,6 ± 0,2 m^2/g (BET) was chosen as a primary initial disperse metal. According to transmission electron microscopy (TEM) the size of particles of aluminum powder varies mainly from a few to tens microns, though there are smaller particles, the size of which varies from 45 to 70 nm. Modes of modification, used in the research, do not change a form and a size of particles. Aluminum powder was modified using different modes at a room temperature in vapors of triamon (T), alkamon (A) and hydrophobic silicone-organic liquid (HSL), which is based on ethylhydridsiloxane. Modes of modification of powder were fully described in prior researches [2, 5].

Composition of the samples was determined by using EDX-spectroscopy (analytical attachment EDAX/TSL, shooting mode – 6 kV) and X-ray fluorescence spectroscopy (XRF, Bruker S4 Explorer). The water-repellent properties of the sample were measured gravimetrically using desiccator method at a relative water vapor pressure p/p_0 = 0,98 ± 0,02. Al powder, modified in vapors of quaternary ammonium compounds and/or HSL, was added to an industrial oil I-20 (lubricant). Force of friction (F_{fr}) and coefficient of friction of the system with oil I-20 with Al-additives was determined by using friction testing machine (DM-29M). The tribological pair was Steel 45 (GOST 1050-88) – Bronze Br Aj 9-4 (GOST 18175-78). Moreover, freshly obtained tribological properties of heterogeneus systems were compared with early obtained integral index of friction (*D*), which is proportional to force of friction. Integral index of friction was measured on certified device ARP-11 (frequency range: 20–300 kHz) using a method of acoustic emission [5].

Samples, based on aluminum powder PAP-2, which were examined, are listed below (except initial Al powder). Al/T – sample with chemisorpted triamon; Al/A – sample with chemisorpted alkamon; Al/(A+T) – sample, treated in vapors of mixture of triamon and alkamon; Al/T/A – sample treated consistently with alkamon and triamon; Al/A/T (changed the consistency of treating) and Al/T/T (2 layers of triamon). In order to examine the influence of number of low molecular triamon underlayers on the properties of external layer of ethylhydridsiloxane (HSL), we synthesized completely new samples (Al/T/HSL, Al/T/T/HSL, Al/T/T/T/HSL). Triamon has a molecular formula of $\left[\left(HOC_2H_4\right)_3 N^{+}CH_3\right]\left[CH_3SO_3^{-}\right]$. The main difference between alkamon and triamon is that alkamon has much bigger radicals (C_{16}–C_{18}) [4].

3. Results and discussion

According to figure 1 system with Al/(A+T) sample has the largest decline of coefficient of friction. Al-additive, which modified in vapors of HSL [6] and which has the highest hydrophobicity [5], does not significantly improve the anti-frictional properties of initial lubricant (see also the characteristics of Al/HSL in table 1). The anti-frictional properties significantly improved with inclusion of 1–2 T-underlayers under HSL-layer. Similar effect was discovered for Al/T/A sample, using the friction testing machine DM-29M (figure 1). It is important to note that Al/A/T additive, with alkamon underlayer does not cause F_{fr} to decrease, as much as Al/T/A does (figure 1, table 1).

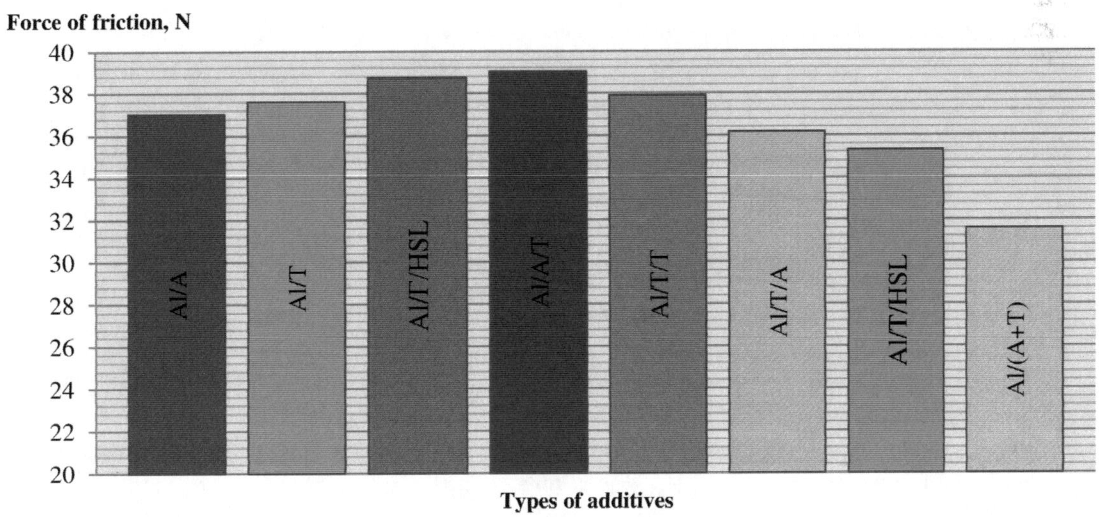

Figure 1. Force of friction in systems with lubricants, containing Al-additives, modified using different modes (the weight load was N = 5000 N; lubricants contained 0,5 % of additives by mass).

Modification of aluminum powder in vapors of a mixture (A+T) gives a greater improvement in anti-frictional properties than samples modified in vapors of alkamon or triamon separately (figure 1) [12].

This synergetic effect can be better noticed by the reduction of integral index of friction using acoustic emission method [5]. Setting for acoustic emission (47 MPa) can produce a pressure almost 3 times greater than a friction machine (17 MPa) can. That explains why the value of *D* in the decrease of friction in the system with Al/(A+T) is sharper than in the figure 2.

Table 1. The influence of type of Al-additive (0,5 % by mass) on $F_{fr} = \Phi(N)$ equation, the change of F_{fr} (ΔF_{fr}) and coefficient of friction (f) in comparison with the initial lubricant (I-20).

№	Al-additive (lubricant)	$F_{fr} = \Phi(N)$ equation	R^2	ΔF_{fr} (average), %	ΔF_{fr} (N = 5 кN), %	f (N = 3,5 кN)
1	Al/(A+T)	$y = 0{,}037x + 12{,}47$	0,991	–11,41	–15,92	0,0075
2	Al/T/HSL	$y = 0{,}043x + 12{,}15$	0,986	–7,42	–5,99	0,0075
3	Al/T/A	$y = 0{,}048x + 10{,}81$	0,992	–7,75	–3,69	0,0079
4	Al/A	$y = 0{,}050x + 12{,}05$	0,997	–1,05	–1,40	0,0089
5	I-20 (no add.)	$y = 0{,}050x + 12{,}29$	0,994	0	0	0,0089
6	Al/T	$y = 0{,}050x + 11{,}86$	0,999	–1,52	0,13	0,0087
7	Al/T/T/HSL	$y = 0{,}049x + 12{,}49$	0,993	–0,21	0,89	0,0085
8	Al/T/T/T/HSL	$y = 0{,}048x + 12{,}50$	0,995	–0,20	0,88	0,0086
9	Al/T/T	$y = 0{,}051x + 11{,}59$	0,994	–1,59	0,89	0,0086
10	Al/HSL	$y = 0{,}048x + 13{,}22$	0,994	2,32	3,80	0,0086
11	Al/A/T	$y = 0{,}050x + 12{,}68$	0,992	1,96	3,95	0,0085
12	Al (PAP-2)	$y = 0{,}065x + 11{,}74$	0,997	12,61	20	0,0101

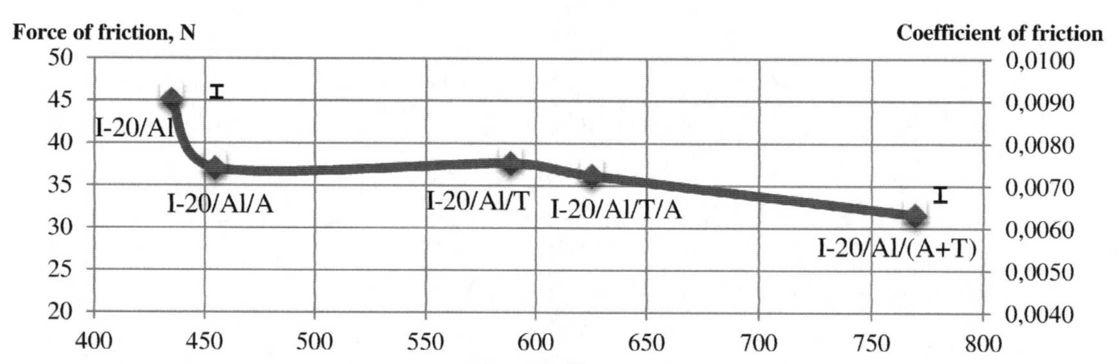

Figure 2. The dependence of F_{fr} and f on $1/a$, where a – adsorption of water on Al-additive (N = 5 кN, P = 17 MPa).

According to the ideas shown in the researches [1, 5, 7], at high load pressure as the liquid lubricant (I-20) is pressed out of the area of contact, system verges in the mode of "dry friction". Thus, anti-frictional properties of the system significantly depend from characteristics of the surface of solid

Journal of Physics: Conference Series **729** (2016) 012026 doi:10.1088/1742-6596/729/1/012026

additives. Therefore, it becomes evident that in the soft friction conditions the effects caused by physical and chemical properties of Al-additives surface (table 1), become less obvious than in the data shown in and researches [1, 5].

The relationships between frictional force and the weight load of the lubricant with Al-additives and of the original oil I-20 (table 1), can be approximated by the linear with the reliability degree R^2 measured between 0,986–0,999. Obtained $y = kx + b$ form of equations are similar to the boundary friction equation in which intermolecular interaction of contacting surfaces is considered [8]:

$$F_{\text{fr}} = k\left(N + Sp_0\right) = kN + kSp_0,$$

where k – friction coefficient in the measuring range N; N – the force of normal pressure; S – contact area between solids; p_0 – additional pressure, caused by intermolecular interaction forces.

Equations with minimal proportionality factors k and the smallest value of R^2 (0,991 and 0,986) correspond to Al-additives (Al/(A+T), Al/T/HSL respectively) which reduce the force of friction the most. Free term of the equation, which is stands for intermolecular interaction, is the smallest for the additive Al/T/A and the biggest for the additive Al/HSL (table 1). Lubricant with Al/T/A ranks among the top three lubricants with the best anti-frictional properties. Otherwise lubricant with Al/HSL has the largest friction coefficient f.

According to the data shown in the table 1, Al-additives with external HSL-layer on the surface may be placed by the ability to reduce F_{fr} in the following way:

$$Al/T/HSL > Al/T/T/HSL > Al/T/T/T/HSL > Al/HSL$$

← Reduction of F_{fr}

Thus, the additive with one T-underlayer improves the anti-frictional properties of lubricant the most. Such additives as Al/T/T/HSL and Al/T/T/T/HSL have a little positive impact on the F_{fr} in the weight load range from 50 to 350 kgf. Moreover, they lower the anti-frictional properties of the oil I-20 (in the table 1 $\Delta F_{\text{fr}} > 0$) at the high weight load (more than 500 kgf).

In the research [9] it was established that the increase in the number of T-underlayers under the external A-layer on the aluminum powder reduces the water-repellent properties of the sample. So, it is possible that the best anti-frictional properties of the sample Al/T/HSL and the increase in hydrophobicity of the sample Al/T/A is just one more example of the monolayer effect, which was derived by Aleskovsky [10]. As a number of T-underlayers increases, the external hydrophobic layer becomes more distanced from the aluminum surface, which leads to the decline of some tribochemical properties of the samples. This fact indicates of influence of force of interaction between the metal surface and the external layer on tribochemical properties of the sample. According to Aleskovsky, after the layering of 3–4 layers the influence of aluminum surface nearly disappears. According to Abramzon [4], hydrophobization of the surface of metal and high level of adhesion between metal and a layer of surfactant are necessary to obtain good anti-frictional and protective properties of the sample. The results of this research confirm that the adhesion factor is a major factor among two factors listed above, which was earlier proved using the acoustic emission method for different metal-fillers [1, 11].

According to the figure 2, the decrease of force of friction in the system is proportional to the decrease of the level of adsorption of the water vapor on the metal. In other words, the force of friction is indirectly proportional to the hydrophobicity of Al-additives.

The similar results were obtained earlier [7]. However, the addition of the water-repellent (A or HSL) on the aluminum does not decrease F_{fr} of the lubricants by much. The most efficient out of all Al-additives are the additives which have external layer of triamon with the small (C_1–C_2) organic substituents on the nitrogen atom. This fact leads to the penetration of the T-molecules to the gaps in the industrially applied stearic nanofilm on the aluminum. We have concluded that the steric availability of the nitrogen atoms in the triamon eases the forming of heteroatomic interaction between nitrogen and metal and between triamon and external layer A or HSL. Factors that are listed above contribute to the stabilization of the external layer of Al-additive and amplification of the anti-frictional effect of the lubricant with this additive.

Comparing the results to the data obtained for oil I-20 with nanopowder of soft metals (Cu, Zn, Pb, Brass) and received using electroexplosive method [13], shows the following. The synthesized Al-additives allow us to form the lubricants with anti-frictional properties and f, similar to those in the research [13]. The unusual fact is that due to the presence of micron size particles original disperse Aluminum PAP-2 cannot be named as nanopowder [11]. The F_{fr} and f decrease as a result of application of a nanofilm of quaternary ammonium compounds or HSL to a metal surface.

4. Conclusions

Thereby, in the circumstances when concentrations of additives, dispersity of solid-additive, mechanical pressure, the nature of tribological pair, and the temperature of the experiment are equal for all samples; it was found that as a number of T-underlayers included in Al-additives with chemisorpted layer of ethylhydridsiloxane reduces from 3 to 1, the force of friction and coefficient of friction also reduce in heterogeneous systems like oil I-20 with Al-additive. The positive impact of the triamon underlayer on the water-repellent properties of Al-additives and tribosystem with these additives is associated with small molecular size of triamon, steric availability of nitrogen atom, which eases the stabilization of system, as well as adsorptional and anti-frictional properties of the external surface of metal.

References

[1] Bystrov D S, Syrkov A G, Pantyushin I V and Vakhreneva T G 2009 *Chemical physics and mezoscopy* **4** 462
[2] Syrkov A G 2013 *Russian Journal of General Chemistry* **8** 1621
[3] Voronchikhina L I, Zhuravlev O E, Andrianova E V and Krotova N I 2014 *International journal of applied and fundamental research* **3** 139
[4] Abramzon A A and Shchukin E D 1984 *Surface Phenomena and Surfactants. Reference Book* (Leningrad: Chemistry)
[5] Syrkov A G, Fadeev D V, Taraban V V and Silivanov M O 2014 *Condensed matter and interphase boundaries* **2** 215
[6] Khanashvili L M and Andrianov K A 1983 *Technology of organoelemented monomers and polymers* (Moscow: Chemistry)
[7] Syrkov A G 2013 *From Laboratory to Industry Smart Nanoobjects* ed K Levine (New York: Nova Science Publishers)
[8] Trofimova T N 2010 *The course of physics* (Moscow: Publishing center "Akademiya")
[9] Bystrov D S 2009 *Nanostructured regulation of anti-friction properties and reactivity of the surface of steel and aluminum* Dis. … cand. chem. sc. SPb: SPbGTI(TU)
[10] Aleskovsky V B 1996 *Chemistry of supramolecular compounds* (Saint-Petersburg: Publishing house of Saint-Petersburg State University)
[11] http://www.icp.ac.ru/conferences/old/Nanochem/Kolesnikova
[12] Syrkov A G, Bystrov D S, Zhurenkova L A and Vahreneva T G 2009 *Russian journal of non-ferrous metals* **2** 79
[13] Tarasov S, Kolubaev A, Belyaev S, Lerner M *et al* 2002 *Wear* **252** 63

Solid lubricant mass contact transfer technology usage for vacuum ball bearings longevity increasing

B Arzymatov and E Deulin

Moscow State Technical University, Russia

E-mail: Arzymatov_bolot@mail.ru, deulin@bmstu.ru

Abstract. A contact mass transfer technological method of solid lubricant deposition on components of vacuum ball bearings is presented. Physics-mathematical model of process contact mass transfer is being considered. The experimental results of ball bearings covered with solid lubricant longevity in vacuum are presented. It is shown that solid lubricant of contact mass transfer method deposition is prospective for ball bearing longevity increasing.

1. Introduction

In practice of engineer all greater application is found by the knots of dry friction, the elements of that are made from "dry" construction materials, or on them hard lubricating coverages (HLC) as that apply different organic and inorganic compounds or thin metallic coverages are inflicted [1]. These coverages, dividing the surfaces of friction, facilitate and accelerate the wear process, diminish the coefficient of friction and eliminate possibility of grasping and jamming of surfaces at a friction. The results based on the use of method of process contact mass transfer (CMT) worked out in Bauman Moscow State Technical University (BMSTU) are presented below [2]. As a method while is small studied, then the task of researches was determination, both technological modes of his realization and longevity of coverage during his work in a vacuum.

2. Theoretical bases

Technological basis of process is the shock causing of HLC on the working (friction) surfaces of knots of mechanism prepared to work in a controlled environment, including, in a vacuum. A shock instrument for realization of process of preparation of surfaces is choose metallic marbles, by a radius, submerged in the powder prepared from the ground up material of the future of HLC, with the size of particles 10–40 μm. Chart of "direct" cooperation of instrument and object with force P at causing of hard lubricating coverages with the initial radius of particles shown in figure 1.

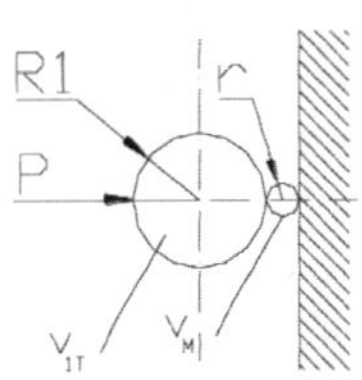

Figure 1. Chart of "direct" shock cooperation of instrument with an object.

Pin pressures in the zone of hitting are determined analytically for the case of "direct" blow of shock body at the wall of working chamber can be written:

$$p_{\max} = \sigma_{\max} = \frac{3}{2}\frac{P}{\pi ab} \qquad (1)$$

where a, b are sizes of semiaxes of elliptic ground of contact.

Experiments and theoretical calculations were created not only for "direct" but also for "sloping" shock co-operating of instrument with an object under different corners. It was shown that pin pressures $p_{\max}$ depending on the parameters of process forming of HLC: of the masses and initial velocities of shock bodies (M_1, V_1), mechanical properties of the inflicted material and shock bodies ($E_M, E_T, \upsilon_M, \upsilon_T$), geometrical sizes of the inflicted material (d, V_M, r) can exceed tensile of material of future friction pair strength, that it talks about plastic introduction of microparticless of HLC in material of the formed friction pair. Experiments on causing of HLC were produced on the stand scheme shown in figure 2.

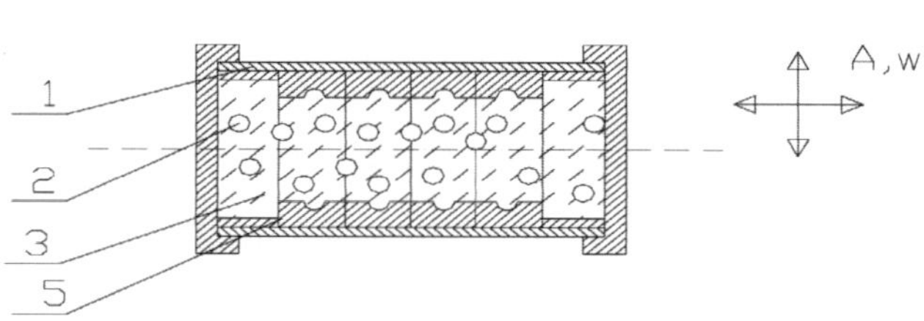

Figure 2. Chart of stand for causing of HLC on the rings of ball-bearings: 1 – working technological chamber; 2 – working bodies (marbles); 3 – inflicted material (HLC as powder); 5 – rings of ball-bearings, subject to causing of HLC. Mark indexes: A – amplitude; W – pulsatance of vibrations of corps of stand.

The different technological stages were prospected CMT, at that the modes of treatment, value of energy, effort, pin tensions, were determined, that allowed to define the basic technological stages. Calculations were produced for the different sizes of amplitudes of vibration at frequency of v = 25 Hz and 70 %-filling with of working oscillation technological chamber material of HLC and processed wares. Were set forth idea about the modes of causing of HLC:

1. Without tearing away mode of oscillation of single mass (shock bodies, wares, inflicted material) at amplitude A: $0 < A^{I} \le A_1 = 0{,}4$ mm;
2. Distribution of mass of HLC, at amplitudes A: $A_1 < A^{II} \le A_2$, where $A_2 = 1{,}3$ mm;
3. Working technological influence at amplitudes $A_2 < A^{III} \le A_3$, where $A_3 = 6$ mm;
4. Exit on the maximum value of destruction of shock bodies at amplitudes: $A_3 < A^{IV}$;
5. Optimal variant of values of amplitudes at frequency of v = 25 Hz are in limits: from 2 to 3 mm.

The key phases of process of causing of HLC were thus distinguished:
– acquisition by the marble of kinetic energy of rebound from the wall of vibrating chamber;
– hitting of two marbles through the particle of material of HLC;
– hitting of marble with the immobile surface of working wall.

3. Results of experimental researches

For the estimation of capacity of vacuum mechanisms with inflicted coverages by the process contact mass transfer next informing parameters were chosen:

1. Outgassing rate, as parameter, characterizing intensity of interchange of gases processes what be going on the surfaces of contact;
2. Moment of resistance of mechanism as parameter, characterizing expenses the energies, realized in a mechanism working in a vacuum.

The results of research of capacity of mechanisms with inflicted HLC are presented in figures 3, 4. From a diagram 2 in figure 3 completion of wear of coverage of MoS is visible attended with growth of moment of friction to 2,7 H·m, it is related to appearance of cohesive cooperation after 21 h of work.

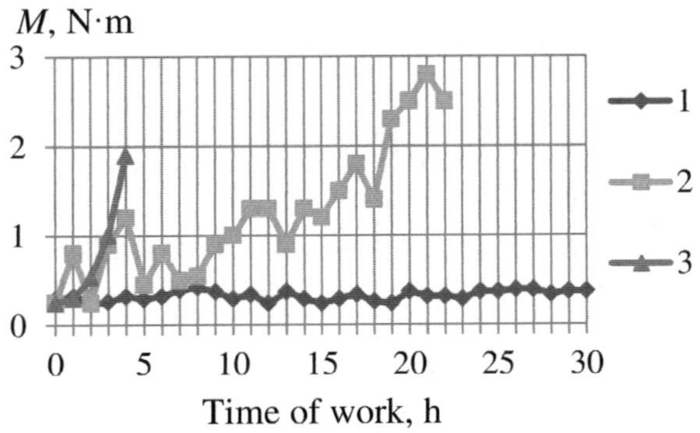

Figure 3. A change of total moment of resistance M (N·m) of two ball-bearings is in the vacuum 10^{-2} Pa. Points on diagrams correspond to coverages: 1 – Pb (lead); 2 – MoS; 3 – without coverage.

From the before published materials [3] we know that appearance of signs of cohesive friction is an index catastrophic wear not only the outgassing rate increases at that but also composition gassed changes with appearance of hydrogen from near surface zones of the particles pulled out at cohesive cooperation.

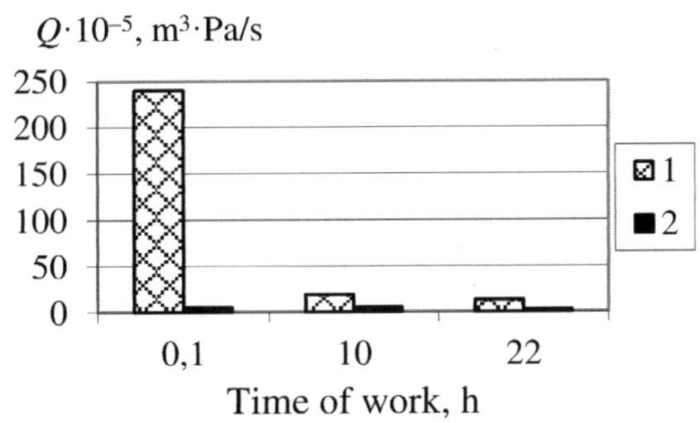

Figure 4. Ball-bearings outgassing flow in vacuum, from the work surfaces covered with MoS: 1 – with ion-plasma method usage; 2 – CMT method usage.

Megascopic outgassing rate during work of ball-bearing of covered MoS in a vacuum as compared to the coverage of Pb (lead), inflicted CMT is explained a method by the change of physical nature of dominant physical factor at a friction. So, enhance able outgassing rate of coverage of MoS in a comparatively small period of work (21,5 h) is explained by moving away of plenty (more than 100) of layers of sorbate from the ground surfaces [4]. The stable and small outgassing rate from coverage of Pb ($Q = 1{,}6\cdot10^{-5}$ $m^3\cdot Pa/s$) is less, than at coverage MoS ($Q = 7\cdot10^{-5}$ $m^3\cdot Pa/s$) talks about different physical nature of interchange of gases, what be going on in the zone of pin cooperation. As leaden coverage stably show the best results on longevity (work is more than 270 000 cycles), at the small moment of resistance (less than 0,2–0,4 H·m) and outgassing rate of $Q = 1{,}6\cdot10^{-5}$ $m^3\cdot Pa/s$, that all of it gives grounds to consider such coverage perspective.

4. Conclusions

Enhanceable outgassing rate in an initial period is explained by moving away of plenty (more than 100) of layers of sorbate from the ground surfaces. The following by it period of the stationary set work corresponds to permanent speed of wear of coverage.

It was shown that completion of wear of coverage is attended with growth of coefficient of friction and temperature, it is related to appearance of cohesive cooperation.

Appearance of signs of cohesive friction is the index of catastrophic wear at that the stream of outgassing rate increases not only but also composition gassed changes with appearance of hydrogen from the near surface zones of the particles pulled out at cohesive cooperation.

The leaden coverage shows the best work results and minimal outgassing flow, that let us consider this coverage perspective.

References

[1] Zeev N A, Kozelkin V V and Gurov A A 1991 *Materials of dry friction, working in a vacuum* (Moscow: Mashinostroenie)

[2] Volchkevich L I and Volchkevich I L 1994 *Metronome* **5–6** 35–7

[3] Deulin E A, Michailov V P, Panfilov Y V and Nevshupa R A 2012 *Mechanics and Physics of Precise Vacuum Mechanisms* (New York: Springer)

[4] Deulin E A, Ikonnikova E I and Tkacheva E V 2013 *Gerald of Bauman Moscow State Technical University* **3** 86–101

Journal of Physics: Conference Series **729** (2016) 012028 doi:10.1088/1742-6596/729/1/012028

TXRF spectrometer on base of the waveguide-resonator with specific design

E Lukianchenko[1], V Egorov[2], V Rudenko[1] and E Egorov[2]

[1] LLC «POLUS», St. Petersburg, Russia
[2] IMT RAS, Chernogolovka, Moscow Distr., Russia

E-mail: emluk@mail.ru

Abstract. This work presents original scheme of the X-ray total reflection spectrometer with the planar X-ray waveguide-resonator (PXWR) with arranged specimen inside of the flux of the waveguide-resonator. Experiment show this X-ray optical scheme gives more effective using the planar waveguide-resonator for purpose of TXRF.

X-ray fluorescence analysis in conditions of the exciting flux external total reflection (TXRF) is one of the effective methods for the material element diagnostics today [1].

The critical parameter of TXRF spectrometry is the radiation density of X-ray exciting beam owing to the beam influence is limited by the total reflection conditions. The enhancing of this parameter can be achieved by increasing of the radiation source power or by raising efficiency of exciting beam formers. Investigations show that the planar X-ray waveguide-resonator (PXWR) is the best device for X-ray beam formation with enhancing radiation density.

Figures 1 and 2 present schemes of the interference fields of X-ray standing waves in extended planar slit clearances with great and nanosize width, accordingly [2].

The total external reflection phenomenon of quasimonochromatic X-ray beam featured by λ_0 wavelength, $\Delta\lambda$ monochromatic degree and $L = \lambda_0^2/\Delta\lambda$ coherence length on the material interface is characterized by the interference field of X-ray standing wave appearance. In conditions of two reflectors placing on same distance of them and forming the planar extended slit clearance it is possible two different situations, in principle. In case of wide slit clearance X-Ray flux will propagate in corresponding with mechanism of the multiple total reflections (figure 1).

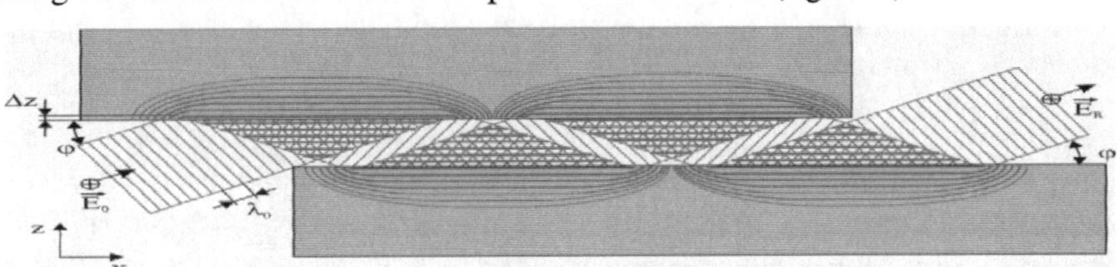

Figure 1. The mechanism of multiple total reflections.

When width of slit clearance will be smaller as half of coherence length local interference areas reach mutual superposition and arise the uniform interference field of X-ray standing wave. This case is corresponding to the mechanism of X-ray flux waveguide-resonance propagation through slit clearance almost without attenuation (figure 2).

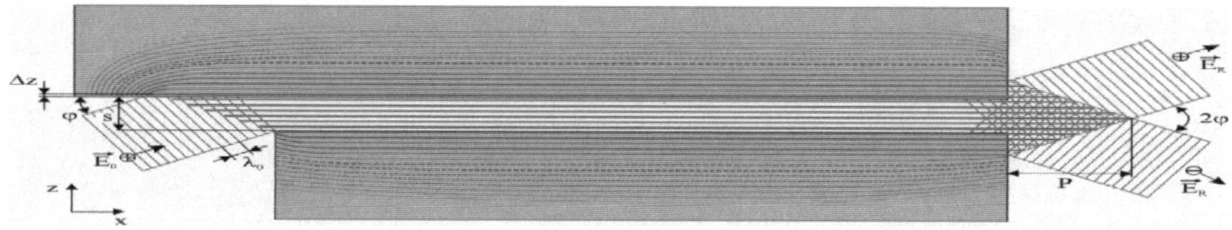

Figure 2. The mechanism of X-ray flux waveguide-resonance propagation through slit clearance almost without attenuation.

PXWR forms X-ray beam with small angular divergence and great radiation density enhancing this parameters for beams formed by any others devices on three orders. At the same time, in our TXRF spectrometer we apply the waveguide-resonator with specific design.

Its construction allows to introduce studied into waveguide-resonance flux propagating in PXWR slit clearance. Such geometry of the sample positioning presents possibility of maximum fluorescence excitation achievement.

Figure 3 demonstrates X-ray optical schemes of TXRF spectrometer built on base of the conventional design (a) and cell of our TXRF spectrometer fabricated on base of PXWR with specific construction provided a studied sample introduction into its waveguide-resonance slit clearance (b).

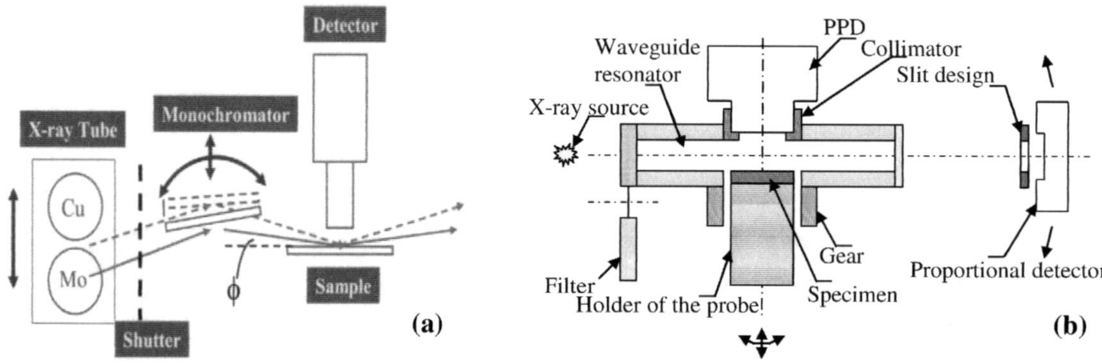

Figure 3. Comparison of X-ray optic scheme of conventional TXRF (a) and with PXWR (b).

X-ray fluorescence excitation in conditions of PXWR specific design application has same advantages in comparison with conventional approach to the excitation:

1. The excitation takes place in the waveguide-resonance flux which provides maximum of the secondary fluorescence yield;
2. There is opportunity for studied target transportation in the flux cross-section;
3. The excitation conditions do not depend from the studied sample aggregate state (solid, liquid, powdery states);
4. PXWR construction allows to change experimental conditions in the measurement process.

The experimental advantages of our TXRF spectrometer (TXRF-PXWR) were confirmed by investigation of dry liquid residues, thin films and coatings, filter object, solid materials, powder-like samples, tissues, blood samples, foodstuffs, mineralogical thin sections and so on.

The analytical characteristics of new investigating setup were studied in conditions of application two X-ray sources BSW27 (Ag) and BSW27 (Mo) in regime $U = 40$ kV, $J = 4$ mA. The measurements were executed using standard specimens prepared on base of dry liquid residues of the elements of middle group [3]. Experiments show that our TXRF-PXWR spectrometer is characterized by pollutions detection limits corresponding to analytical interval $(1–5)10^{-13}$ g. At the same time it is known the conventional TXRF spectrometer are limited by the detection limits near $(1–3)10^{-12}$ g [1].

Advantages of TXRF study of object inside of the waveguide-resonance flux can be demonstrated on base of results obtained for the Ta coating (t = 10 nm) on glassceramic substrate. Figures 4 shows

TXRF spectra collected in conditions of pure surface geometry (a) and in geometry of the sample introduction into the waveguide-resonance channel on depth $l = 20$ nm (b).

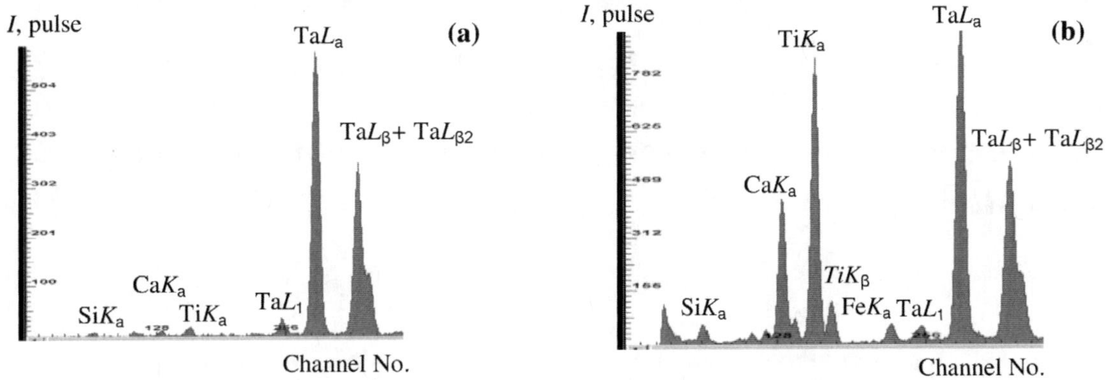

Figure 4. TXRF spectra collected in conditions of pure surface geometry (a) and in geometry of the sample introduction into the waveguide-resonance channel (b).

In the first case (Figure 4(a)) TXRF spectrum contains lines of the coating only. Other spectrum (Figure 4(b)) consist of Ta lines and lines of atoms reflected the substrate composition. Intensity yield of Ta lines are equivalent on the both spectra.

Appearing of lines correlated to substrate atoms presents opportunities to study element composition of layers which is characterized by deep bedding. The aggregate possibilities and set of advantages appearing at using of PXWR with specific construction for practice TXRF measurements application allow to announce about creation of new experimental procedure in frame of TXRF material study.

Acknowledgments
The investigations were partial financially supported by Russian Foundation for Basic Research (project 15-8-02618).

References

[1] Klokenkamper R and von Bohlen A 2015 *Total reflection X-ray fluorescence analysis and related methods* (New Jersey: Willey)
[2] Egorov V K and Egorov E V 2004 *Spectrochim. Acta B* **59** 1049–69
[3] Lukianchenko E M, Egorov V K, Rudenko V N and Egorov E V 2013 *II Russian analytical congress (Moscow)* p 98

Journal of Physics: Conference Series **729** (2016) 012029 doi:10.1088/1742-6596/729/1/012029

The energy of the vacuum related to the theory of energy absorption

A P Danilov

Institute of Problems of Sulfur and Mineral Fuels, Moscow, Russia

E-mail: volinad@mail.ru

Abstract. The primary objective in this article is to investigate a new source of renewable energy, the existence of the vacuum in five environments, and the possibility of absorption of the explosion. The study has also addressed the development of new principles in the motor industry, protection against explosions, new principles of mineral processing and new types of explosives. Also, this study may offer some insight into new approaches in solving problems in thermodynamics, the development of gravity waves, the basis of renewable energy source, and the mechanism of the emergence of gravitational waves.

Everything is relative in this world. Also, the potential and kinetic energy of the body can be absorbed meeting barrier, reducing the rate to zero and granulating the body part. Now this concept does not exist. But the discrepancy of the Laws of Newton and Einstein must be overcome. I do not seem to sound appealing, do I? However, energy absorption of the vacuum certainly deserves attention. The nature of the vacuum is known in the Universe, in air and water environments.

The interaction of bodies is characterized by the potential energy. Kinetic is the movement of the body. And potential and kinetic energy change as a result of the interaction of bodies, whose acting on the body forces perform work that is different from zero.

If we consider the energy changes in the interactions of bodies, forming a closed system, several bodies interacting only by gravitational forces and elastic forces and no external forces do not act on them, if there are any interactions of the bodies of work of elastic forces or gravitational forces equal the change in potential energy of the bodies, taken with the opposite sign. The energy of a moving body depends on its speed.

From the formula of Einstein

$$E = mc^2,$$

you can come to the conclusion that energy and body mass are one, and also confirms that energy is a scalar characteristic of matter in motion.

Vacuum is a special medium material, which is not inherent in the properties and peculiarities of existence of matter, but energy absorption of vacuum has tremendous strength that is comparable to a thermonuclear reaction. To do this, fair formula:

$$E = mv^2,$$

where v is the absorption rate of vacuum close to the speed of propagation of electromagnetic waves must also be constant for each environment (gas, liquid, solid rock, fire, and the actual vacuum with

Journal of Physics: Conference Series **729** (2016) 012029 doi:10.1088/1742-6596/729/1/012029

different density of the vacuum in outer space). The speed of propagation of electromagnetic wave is equal to the known constant of 299 792 458 ± 1.2 m/s.

It was the accidents at the Sayano-Shushenskaya hydroelectric power station (figures 1, 2) and the mine Raspadskaya that helped create the theory of energy absorption because I personally took direct or indirect part in the investigation. The safety of the miners was in the first place.

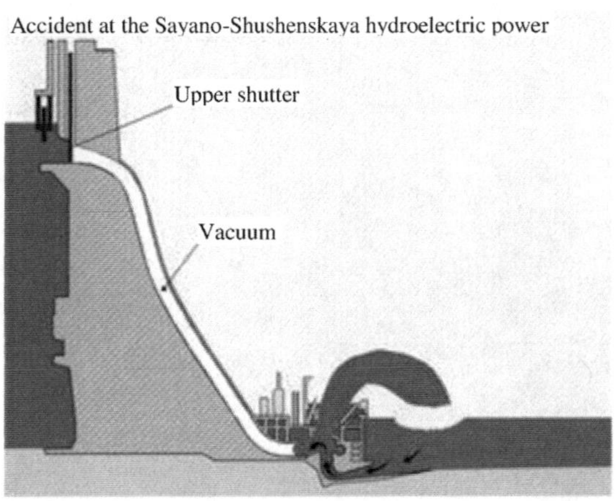

Figure 1. The time of the surge.

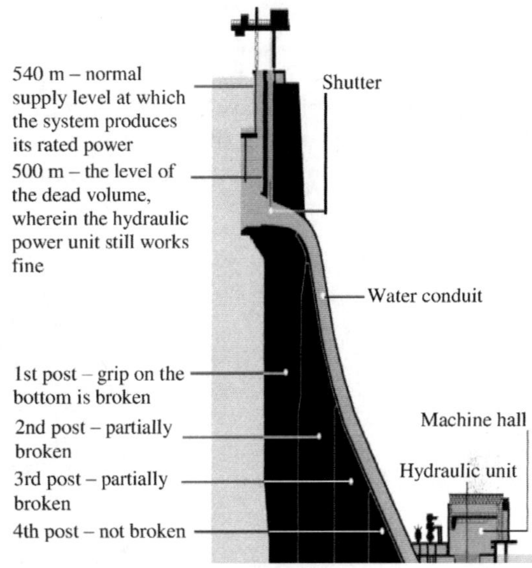

Figure 2. The scheme of destruction after the surge.

In the strata of the Earth. On the Raspadskaya mine, when the accident occurred avalanche emission of methane, there was working in the mine a harvester company "Joy", with a capacity of 1 000 t/h. As is known, methane at a certain concentration can ignite itself. After the explosion there was a shift of the working layer, the vacuum formed a cavity, is initialized and exploded.

Mixed with methane, where the concentration has reached the explosive state, followed by rock burst (the second explosion), resulting in the creation of fractured area on the seam and copious methane (shale gas technology). Methane is explosive at concentrations from 4.4 to 17 %.

Sir Humphry Davy (chemist) in 1813, concluded from their analysis that the mine gas is a mixture of methane CH_4 with a small amount of nitrogen N_2 and carbon dioxide CO_2, i. e. that it is qualitatively identical in composition to the gas emitted from swamps.

What is the oxidation? Uncovered through the formation surrounding the incision Raspadsky. Again a controlled situation is not achieved, and there occurs the third explosion of methane–coal–dust mixture (so-called vacuum explosion, which burns out the oxygen). All explosions create a vacuum space. Due to the expansion of vacuum space, shock wave occur (figure 3).

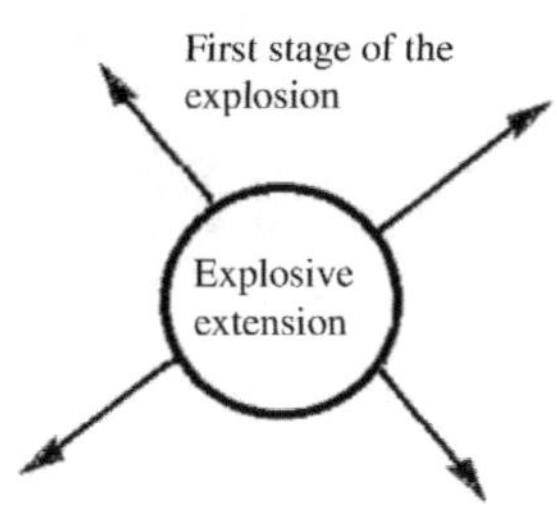

Figure 3. Area expansion during the explosion.

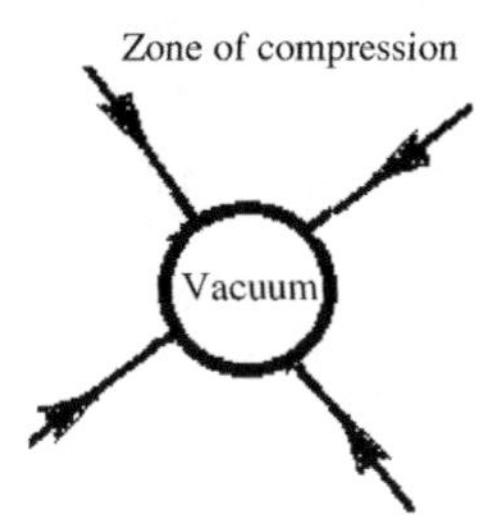

Figure 4. A zone of compression after the explosion.

Journal of Physics: Conference Series **729** (2016) 012029 doi:10.1088/1742-6596/729/1/012029

Wave process supports the reduction of the vacuum, there is a pulse of shock waves. With the explosion of dust–air mixture, the volume of the vacuum burnout much higher point of explosion, therefore, the wave of destruction is much stronger. Quite well shows the effects of wave motion in the works of academician V Adushkin.

If you remember the test of a hydrogen bomb and the description in the papers that the shock wave circled 4 times the globe. How would accumulative the explosion of the directed action. It was pulsing shell of the Earth, which at its stretching could simply collapse and be absorbed by the vacuum of space (figure 4). It was clear to academician A Sakharov, member of the creation of the hydrogen bomb. On this principle black holes are formed.

After the earthquakes in the strata of the Earth, there is a displacement of the monolithic slabs in areas of faults, forming a vacuum cavity (figure 5).

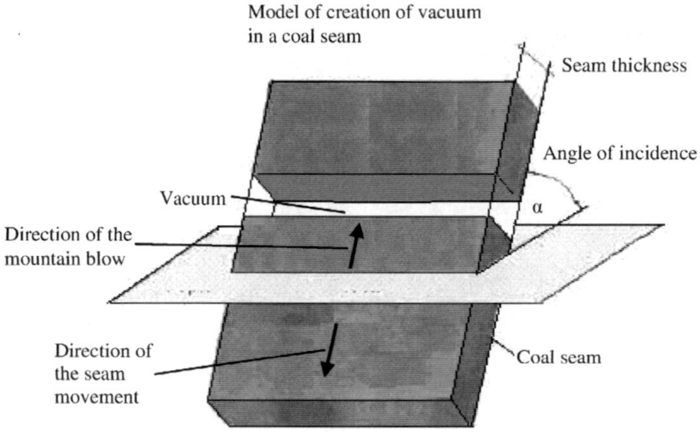

Figure 5. How vacuum emerges in the strata of the Earth.

A vacuum cavity occurs when the displacement of rock slabs, especially dangerous in steeply inclined coal seams, when the concentration of methane reaches a very dangerous level.

In the air. With air flow, with temperature difference, turbulence arises – an appendage vacuum. In the zone of turbulence, jets starts to shake, causing the flutter, and turboprop begin to fall in the air holes. A good example, perhaps, happened with the Malaysian Boeing over the Atlantic. In vacuum space there was a blow of air masses, where the plane, which instantly collapsed?

The presence of the vacuum spaces is well characterized by the Bermuda triangle when there are many ocean currents and the movement of air masses and creates vacuum cavity that swallows ships and aircraft.

In the fire. During combustion, especially high-rise buildings, due to the temperature difference between the combustion of different composite materials, are vacuum gaps, which are due to wave dynamics (Adushkin V) to spread the fire fast speeds.

Especially dangerous crown fire in the forest that creates the wave motion of the fire, where the hot air and the top layer with a lowered temperature creates a wind effect from the vacuum pulses.

In space (vacuum). In the vacuum of space vacuum observed layers with different degrees of sparseness (density), which create black holes. Also a black hole absorbs a galaxy and the planet with the violation of the energy balance thereof, with a damaged shell, which will allow previously inaccessible to study the effects of strong gravity.

With the destruction of the ionosphere of the Earth by a nuclear explosion all the living and the dead can be placed freely in the newly formed black hole.

This is a hypothesis, but they deserve attention, in the present state of science, we learned to blow up, but they can't absorb explosions, save the shell of the Earth.

Journal of Physics: Conference Series **729** (2016) 012029 doi:10.1088/1742-6596/729/1/012029

Independently, staff has developed vacuum-air engine, instead of internal combustion engine, where one cylinder receives vacuum and the other compressed air from receiver. Not allocated at the same temperature. The efficiency of the engine after manufacture, to be determined.

In space such an engine to operate easier, it does not need vacuum pump and receivers. Such an engine generator can provide the vital functions of the spacecraft (constrained by the presence of compressed air (gas) or circulated fluid – the heavier the environment in which a vacuum is created, the higher energy absorption, whether it be air, helium, water, transformer oil. Theoretically, the vacuum can be created in any gas, if in nitrogen, such grenades can extinguish the fire if the methane, it can explode. All this is already applied in practice.

Vacuum – renewable source of energy, comparable to nuclear fusion, which is supposed to curb!

Working on the theory of energy absorption, a vacuum unit and motors have been invented. Chipper vacuum for the purpose of adsorption was invented, in Irkutsk, acted in the Institute, which bankrupted the author died.

The copyright certificate on the vacuum 4-stroke engine, which schematically, excluding the mass of the piston was designed for the idea in Kazakhstan.

Vacuum pneumatic tool, high destructive power and crusher have specific settings that allow you easier to destroy rocks. Information: in mineral processing, crushing energy ranges from 30 to 90 %, depending on the strength of the rocks. It should be clarified that metronome coal, during crushing and extraction of moisture (adsorption) can create a threat of explosion concentration of methane.

This gives new opportunity to the development of the industrial potential of the vacuum community!

Theory of energy absorption should be focused on the development and implementation (this is not a dogma):

– new principles of enriching of coal;

– study and development of the theory and practice of physical and technical vacuum, according to the theory of energy Absorption;

– personal protection for the rescuers and miners;

– downhole and crosscut proof installations;

– absorbers of the explosion in the terrorist attacks;

– vacuum filter fire;

– air-vacuum engine;

– liquid-vacuum engine;

– rotary vacuum crushing;

– excess speed movement principles of underwater objects;

– new mechanisms and systems for destruction of rocks;

– new technologies and the principles of enrichment of minerals and extract them from the poor rocks and tailings of processing plants;

– new types of industrial explosives;

– transfer of coal into a higher quality;

– new methods and blasting methods.

This is not a complete list of everything new and perfect that you can perform and produce on the basis of the theory of energy absorption.

The article talks about a new source of renewable energy, the existence of the vacuum in five environments, the possibility of absorption of the explosion, the development of new principles in motor engineering, protection against explosions, new principles of mineral dressing, new explosives, new approaches to solving problems in thermodynamics and the development of the gravitational wave, the change of the basic energy carriers.

References

[1] Lorentz H A, Poincare A, Einstein A and Minkowski H 1935 *The principle of relativity. The collection of works of classics of relativism* (Moscow–Leningrad: ONTI)

[2] Einstein A 1905 *Annalen der Physik*
[3] Poincare A 1974 *Selected works* (Moscow: Nauka)
[4] Adushkin V V and Spivak A A 1993 *Geomechanics of large-scale explosions*
[5] Ganopol'skii I M, Baron V L, Pupkov V V and Water V I 2007 *Methods of blasting. Special blasting operations* (Moscow: Moscow State Mining University)
[6] Danilov A P 2014 *Coal*
[7] Danilov A P *Vacuum energy, on the theory of energy absorption*
[8] Kantowicz A and Geopano N 1989 *Mining machine*
[9] Nesterov S B and Belyaev E V 2015 *Vacuum technology* (Moscow: Novella)

Journal of Physics: Conference Series **729** (2016) 012030 doi:10.1088/1742-6596/729/1/012030

Analysis of the plasma radiation spectra with lines of significantly varying intensity

D K Kostrin[1], A A Lisenkov[2], A A Uhov[1] and A N Ramazanov[1]

[1] Department of electronic instruments and devices, Saint Petersburg Electrotechnical University "LETI", 197376, Saint Petersburg, Russia
[2] Institute of Problems of Mechanical Engineering, Russian Academy of Sciences, 199178, Saint Petersburg, Russia

E-mail: dkkostrin@mail.ru

Abstract. Range of charge accumulation times for the charge-coupled device photodetector, in which there is linearity of its signal characteristics, is demonstrated. The influence of the blooming effect during the saturation of a signal on the form of spectral lines is shown. Method of simultaneous analysis of lines of significantly different intensity using multiple summations of the emission spectra, obtained with a small time of charge accumulation, is proposed.

1. Introduction

Modern optical spectrometers are constructed on the basis of charge-coupled device (CCD) photodetectors [1]. The choice of the CCD for the construction of the spectrometer optical part is not only due to its high sensitivity, this detector has a number of other advantages. Charge emerging in the element of the CCD storage sections is proportional to the registered radiation intensity and the time interval during which the flow of charge into the potential well is blocked by the shutter.

The effect of interchangeability of radiation intensity and charge accumulation time creates possibility to control the sensitivity of the spectral device using software. Such method allows detecting the signal from sources with intensities of radiation, which differ in dozens and even hundreds of times.

The necessary duration of charge accumulation in the CCD structure is correlated with the intensity of radiation in the considered spectral range. The lower the level, the longer must be the period of charge accumulation for obtaining acceptable signal amplitude at the output of the photodetector. However, it should be noted that the generation of minority carriers in a semiconductor can occur not only due to the photoelectric effect, but also as a result of thermal ionization. Thus, for each semiconductor structure at a certain temperature there is a limit to the duration of charge accumulation and accordingly the minimum (threshold) energy of radiation that can be registered on the background of generation of thermal electrons and noise of different nature.

The number n of the thermal electrons generated in the CCD cell for time t can be written as:

$$n \approx \frac{StT^2}{q}\exp\left(-\frac{\Delta E}{kT}\right), \tag{1}$$

where S – is the cell area; T – CCD crystal temperature; q – electron charge; ΔE – band gap of the semiconductor; k – Boltzmann constant.

The dark current is a result of spontaneous generation of electron-hole pairs. One of the most effective ways to reduce dark current and thermal noise is cooling of the CCD crystal, which in turn allows lowering the threshold energy of the registered radiation. In devices that do not require very low values of dark current Peltier elements are typically used for thermoelectric cooling.

2. Research of a light-signal characteristic of the CCD used in spectrometric equipment

Let us consider a light-signal characteristic of the TCD1304 CCD used as a receiver in many modern compact spectrometers, in particular in developed at the Department of electronic instruments and devices of Saint-Petersburg Electrotechnical University "LETI" spectrometric complex ISM3600 [2]. Time of charge accumulation in the considered spectrometer can be set in the range of 0.02...5000 ms. The lower limit is determined by the required minimum time of charge accumulation, and upper limit is due to the increase of dark current to the level of the maximum signal.

Figure 1 shows a light-signal characteristic of the CCD, obtained by measuring the spectrum of radiation of fixed intensity source – halogen incandescent lamp, connected into the current stabilizing circuit. In this case, the dark current of the CCD is subtracted from the total signal by software.

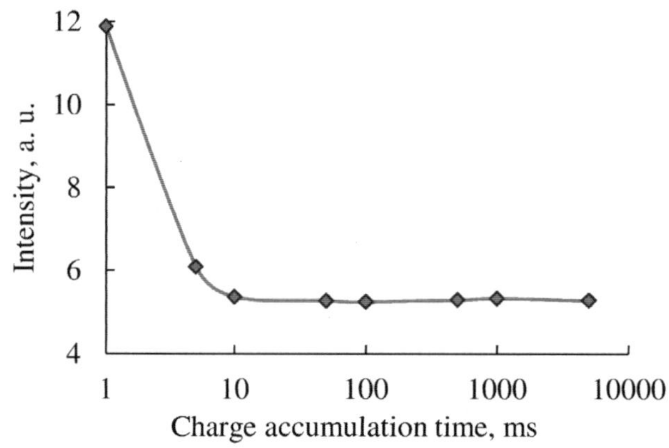

Figure 1. Light-signal characteristic of the TCD1304 photodetector.

The graph (figure 1) shows linearity of the characteristic in a wide range of charge accumulation time – from 10 ms up to 5 s. There is a combination of measuring ranges – reduction of the input signal can be compensated by a proportional increase in the charge accumulation time. However, in the region of small charge accumulation times the linearity is distorted, which may be due to the properties of the CCD and the impact of the device digital circuitry. Thus, it is possible to recommend carrying out measurements of the emission spectra with the charge accumulation time of at least 10 ms. Also increasing the charge accumulation time leads to a decrease in the influence of pulsations of the radiation source, which improves the overall quality of the obtained spectra.

Optical radiation even with a very weak intensity produces electrons in the cells of the storage sections of the CCD photodetector. However the capacity of each pixel of the CCD is not limitless. If the number of photons is sufficiently large, then the quantity of generated charges may be greater than the capacitance of the pixel. In this case the signal reaches saturation and the excess charge starts to flow into adjacent cells – blooming effect occurs. The influence of blooming effect on the output signal of the CCD is reflected in the broadening of the intense lines (figure 2). It is obvious that up to a certain point the increase of the charge accumulation time leads to a linear increase in the intensity of the signal, further the effect of blooming occurs – a significant distortion of the waveform in the region of larger wavelengths can be seen [3].

It can be assumed that this distortion occurs when transferring the charge packet – capacity of cells of the transfer section is less than capacity of CCD pixels. In this case charge that will not fit into the cells of the transfer section when the transportation begins flows into the nearest cells in the direction of motion of the charge packet.

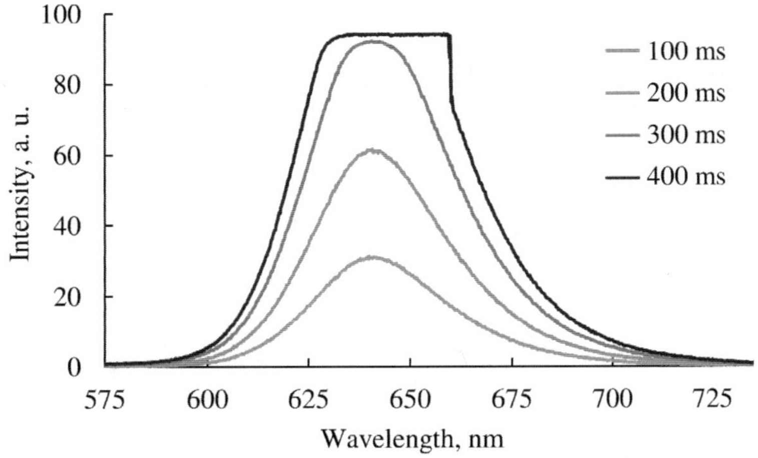

Figure 2. Blooming effect on the example of a continuous spectrum.

It can be concluded that it is necessary to carefully select the charge accumulation time in the CCD in order to avoid the effect of blooming, which could be programmatically processed in the case of uniform broadening, unfortunately not seen in the application of the TCD1304 photodetector because of its technological features.

3. Analysis of the linear spectra with the lines of significantly varying intensity

Let us consider the effect of blooming on the example of a spectrum fragment of a mercury low pressure lamp (figure 3). It can be seen that there is a significant broadening of spectral lines, which leads to considerable difficulties in determining the wavelength of the respective line. It should be noted that in the case of a linear spectrum is also observed the broadening of spectral lines towards the larger wavelengths, i. e. in the direction of charge packet transfer.

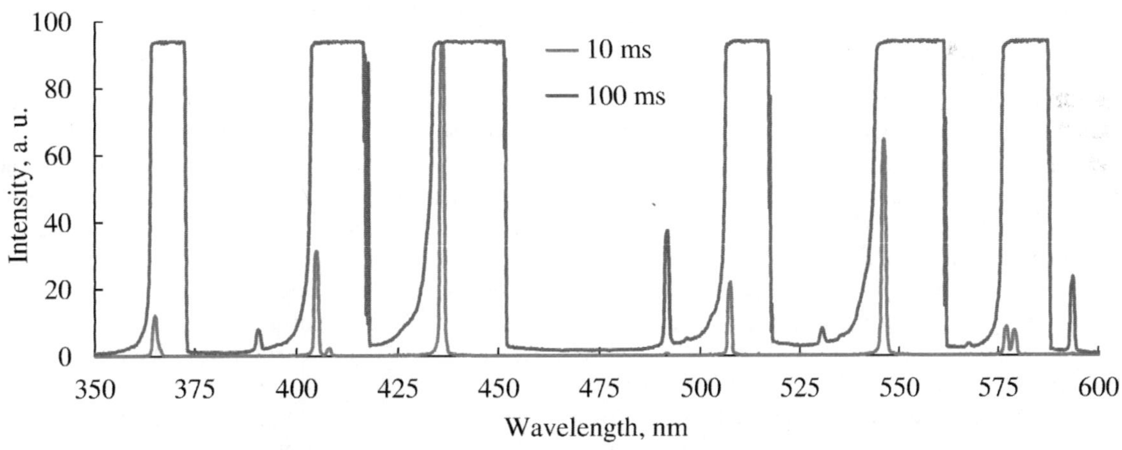

Figure 3. Emission spectra of a mercury lamp at the charge accumulation times of 10 and 100 ms.

Using this spectrum to analyze the composition of a gas mixture [4] is not practically possible because of the significant changes in the shape of spectral lines. In particular the blooming effect leads to unification of the lines of doublets (e. g. doublet 577–579 nm) and triplets into one wide line.

On the other hand, a significant increase in the charge accumulation time resulted in appearance of new spectral lines. The intensity of the individual spectral lines are often extremely small and one reading of the contents of storage sections does not allow to distinguish this weak signal on the background noise of the CCD.

Using multiple summation of a signal with subsequent averaging it is possible to make such lines suitable for analysis. Figure 4 shows a fragment of the emission spectrum of a mercury lamp obtained at the charge accumulation time 10 ms and the sum of ten consistently received such spectra. On last spectrum lines with wavelengths 489 and 491.6 nm earlier merged into one became separate.

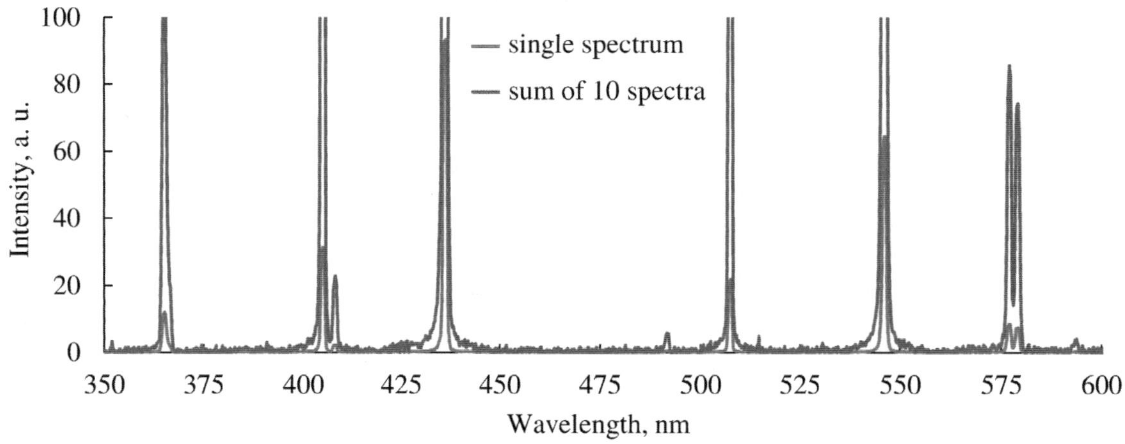

Figure 4. Emission spectrum at the charge accumulation time 10 ms and the sum of 10 such spectra.

It is obvious that the summation of the spectra obtained using a small charge accumulation time on one hand makes it possible to consider spectral lines of low intensity and on another – to get rid of the blooming effect and not to lose information about lines of high intensity.

4. Automatic search of the spectral lines position in the analysis of spectrums

Analysis of the plasma emission spectra may involve the use of algorithms for automatic determination of the position of spectral lines, which allows qualitative and quantitative analysis of the plasma composition [5]. The main requirement is that the search algorithm is high speed, allowing the search of lines in real time.

The task of identifying spectral lines in the linear spectra is necessary for the correct operation of the algorithms of qualitative spectral analysis. There are a number of algorithms used in chemical studies to identify series of peaks and matching them with the profiles of various materials [6]. Most of these algorithms are specialized on the analysis of the peptide components and are not suitable for the research of optical spectra. Methods of detecting peaks in the optical range are not so common. The algorithm to identify the spectral lines must provide cutoff of the noise component, accounting for the asymmetry of a signal and the solution of the problem of the nested peaks, doublets and triplets.

There are a number of approaches to the identification of spectral lines. It is worth noting the following algorithms: finding a local maximum, the method of fitting functions and wavelet analysis. Some of these algorithms do a good job of highlighting the peaks, but they run not fast enough, even using modern computing power, and lose their relevance in the analysis of spectra in real time. Let us consider the most simple and thus fast algorithms to determine the position of the spectral lines.

The first algorithm searches for the position of spectral lines by calculating the so-called center of gravity of the shape. The order of steps in this algorithm is as follows: search of the peaks in the whole spectral range, removing from consideration peaks that have low intensity, search for lower boundaries of the peaks, calculation of the center of gravity for each of the peaks.

This algorithm has sufficient performance and allows to effectively determine the actual position of spectral lines. A significant drawback of the algorithm is the inability to define the position of the lines of an incompletely resolved doublet.

The solution of this problem requires fitting into the shape of a bounding peak two mathematical Gauss, Lorentz or Voigt functions and find their parameters to obtain the existing form of peak. The wavelengths corresponding to the spectral peaks in this case are determined by the parameters of these

Journal of Physics: Conference Series **729** (2016) 012030 doi:10.1088/1742-6596/729/1/012030

functions. This addition to the algorithm largely reduces its performance and is not applicable for analysis in real time.

The second considered algorithm is easier to implement and has higher performance. The procedure for it is as follows: search of the peaks in the whole spectral range, removing from consideration peaks that have low intensity, search for the left and right border of the shape of the peak on half of its height, calculation of the center of the figure.

This algorithm unlike the previous one allows handling partially resolved doublet lines. In the case of having them in spectrum the algorithm moves from a half height of the peak to the intensity of the point corresponding to the separation of adjoined lines.

The results of determining the position of spectral lines for the low pressure mercury lamp with the use of the both considered algorithms are given in table 1. Also shown for comparison are the true values of the wavelengths, as well as the positions of the maxima for the lines in the spectrum.

Table 1. The results of determining the position of spectral lines.

Algorithm	Wavelength of the spectral line, nm						
Maximum of the spectral peak	313.60	365.35	404.00	435.10	545.00	576.15	577.90
Center of the gravity of the shape	313.45	365.45	404.55	435.35	545.30	577.25	577.25
Center of the peak on half height	313.30	365.45	404.65	435.45	545.35	576.85	577.90
Theoretical wavelength value	312.60	365.00	404.67	435.83	546.07	576.96	579.07

Analysis of the data given in table 1 makes it possible to conclude that both algorithms allow obtaining the values of the position of spectral lines sufficiently close to the theoretical values. The accuracy of the second algorithm (despite the fact that it is easier and faster) is higher.

5. Conclusions

Based on the conducted experiments it can be recommended when using spectral equipment built with the use of the TCD1304 photodetector to obtain emission spectra for further analysis when the output signal of the device lies in the range 20...90 a. u. (for the units of the output signal the percentage level from the maximum signal of the CCD which corresponds to its saturation is taken).

The choice of this range is due to the fact that at low values of the signal level significantly decreases the accuracy of analog-to-digital converter included in the conversion circuitry of the device, besides the signal becomes comparable with the noise level, which leads to considerable inaccuracies of the measurement. When the values of the output signal are over 90 a. u. the blooming effect begins to affect the results, leading to a dramatic broadening of the spectral lines, which creates significant difficulties in defining their position.

It can be noted that the algorithms of the search of the peak positions with their simplicity and speed enable to effectively determine the wavelengths of the spectral lines. The obtained data can further be used to evaluate the plasma composition in real time, which allows in the control of technological processes in a timely manner to respond to the changes in the plasma flux composition.

References

[1] Uhov A A, Gerasimov V A, Kostrin D K and Selivanov L M 2014 *Journal of Physics: Conference Series* **567** 012039
[2] Kostrin D K and Oukhov A A 2013 *Biotechnosphere* **3** 21–5
[3] Kostrin D K and Uhov A A 2013 *Testing. Diagnostics* **5** 40–2
[4] Kostrin D K, Uhov A A and Lisenkov A A 2016 *Vakuum in Forschung und Praxis* **3** 34–7
[5] Kostrin D K 2015 *Testing. Diagnostics* **5** 47–50
[6] Vu T N and Laukens K 2013 *Metabolites* **3** 259–76